SOCIAL MEDIA ANALYTICS AND PRACTICAL APPLICATIONS

Emerging Operations Research Methodologies and Applications

Series Editors
Natarajan Gautam
Texas A&M, College Station, USA

A. Ravi Ravindran
The Pennsylvania State University, University Park, USA

Multiple Objective Analytics for Criminal Justice Systems
Gerald W. Evans

Design and Analysis of Closed-Loop Supply Chain Networks
Subramanian Pazhani

Social Media Analytics and Practical Applications
The Change to the Competition Landscape
Subodha Kumar and Liangfei Qiu

For more information about this series, please visit: https://www.routledge.com/ Emerging-Operations-Research-Methodologies-and-Applications/book-series/ CRCEORMA

SOCIAL MEDIA ANALYTICS AND PRACTICAL APPLICATIONS

The Change to the Competition Landscape

Subodha Kumar and Liangfei Qiu

CRC Press
Taylor & Francis Group
Boca Raton London New York

CRC Press is an imprint of the
Taylor & Francis Group, an **informa** business

First edition published 2022
by CRC Press
6000 Broken Sound Parkway NW, Suite 300, Boca Raton, FL
33487-2742
and by CRC Press
2 Park Square, Milton Park, Abingdon, Oxon, OX14 4RN

© 2022 Subodha Kumar and Liangfei Qiu

CRC Press is an imprint of Taylor & Francis Group, LLC

Library of Congress Cataloguing-in-Publication Data
Names: Kumar, Subodha, author. | Qiu, Liangfei, author.
Title: Social media analytics and practical applications : the change to
the competition landscape / Subodha Kumar and Liangfei Qiu.
Description: First edition. | Boca Raton : CRC Press, 2022. | Includes
bibliographical references and index.
Identifiers: LCCN 2021036976 (print) | LCCN 2021036977 (ebook) |
ISBN 9781032051390 (hardback) | ISBN 9781032051406 (paperback)
| ISBN 9781003196198 (ebook)
Subjects: LCSH: Social media and society. | Social media--Influence. |
Business--Data processing.
Classification: LCC HM742 .K86 2022 (print) | LCC HM742 (ebook)
| DDC 302.23/1--dc23
LC record available at https://lccn.loc.gov/2021036976
LC ebook record available at https://lccn.loc.gov/2021036977

ISBN: 978-1-032-05139-0 (hbk)
ISBN: 978-1-032-05140-6 (pbk)
ISBN: 978-1-003-19619-8 (ebk)

DOI: 10.1201/9781003196198

Typeset in Times
by MPS Limited, Dehradun

This book is dedicated to:
my children Medha and Aarna,
my wife Susmita, and
my parents
Subodha Kumar

This book is dedicated to:
my children Adan and Alex,
my wife Bei, and
my parents
Liangfei Qiu

Contents

Preface   ix
Authors   xi

1 How Social Media Shapes the Way We Think   **1**
  1.1   Introduction: "Bellwether"   1
  1.2   How Social Media Shapes the Way Companies and
       Consumers Think   2
  References   3

2 Wisdom of Crowds Meets Social Media   **5**
  2.1   Introduction: Guessing Weight Competition in the
       Age of Social Media   5
  2.2   Public Prediction Markets and Social Media   8
  2.3   Corporate Prediction Markets and Social Media   13
  2.4   Free-Rider Effect and Social Technologies in
       Prediction Markets   16
  References   18

3 Social Media and Firm Strategies   **21**
  3.1   Introduction: Reddit Trading Frenzy and United
       Airlines' Brand Crisis   21
  3.2   Online Management Responses in Social Media   22
  3.3   Firm Strategies and Social Media Platforms   26
      3.3.1   Social Media and User Engagement   26
      3.3.2   Social Media and Pricing Strategies   29
      3.3.3   Social Media and Transactions in
           Marketplaces   30
  3.4   Location-Based Social Media   32
  References   34

4 Social Media, Fake Reviews, and Machine Learning Method   **37**
   4.1   Introduction: Fake News and Fake Reviews   37
   4.2   Fake Review Detection in Social Media   39
   4.3   Sentiment Manipulation as an Operations Activity in
           Social Media   41
   References   45

5 Social Media and Healthcare   **49**
   5.1   Introduction: Online Health Communities   49
   5.2   Online Physician Responses and Social Media   50
   5.3   Online Consultation and Offline Appointments   52
   References   53

6 Social Media and Telecommunications   **55**
   6.1   Introduction: Social Media and 5G   55
   6.2   Sponsored Data Program   56
   6.3   Reward Tasks and Social Media   58
   References   59

7 Future Trends and Challenges in Social Media   **61**
   7.1   Social Trading and Fintech: Individual Investors in the
           Age of Social Media   61
   7.2   Social Media Text, Network, Images, and Videos
           as Data   62
   7.3   Real-Time Nature of Social Media in Operations   63
   7.4   Path Forward   64
   References   65

Preface

Recent years have witnessed an unprecedented explosion in social media (e.g., Facebook, Twitter, Instagram, and Pinterest). Social media is not about changing a specific industry, but rather revolutionizing the nature of how businesses operate and changing the landscape of industry compelition in online retailing, healthcare, telecommunications, and others.

This book aims to provide a framework to understand and analyze the impact of social media in various industries using analytics. This book begins with looking at how social media affects the way we think (Chapter 1). This is followed by the discussion of how social media impacts the wisdom of crowds (Chapter 2). Corporate practitioners have already recognized the power of harnessing the collective wisdom in managing demand risk in supply chains, monitoring leading business indicators, and gathering new ideas and business innovations. As social media technologies have grown explosively, we focus on how these technologies can fundamentally impact the wisdom of crowds. In Chapter 3, we explore how firms can take advantage of the recent advancements in social media technologies and revolutionize the nature of business operations. Next, we examine fake review detection and the use of machine learning methods in social media (Chapter 4). Social media sites suffer from opinion spam and fake reviews. The consequence of such fake reviews is the deterioration of information quality as well as loss in consumer welfare. Chapter 5 investigates social media healthcare platforms, which are rapidly being embraced by patients and physicians. Patients talk to each other and seek expert opinions on these platforms. Physicians use the free service as a self-branding technique to build their reputations. Chapter 6 focuses on the relationship between social media and the upcoming next-generation 5G networks. Telecommunications industry practitioners are investigating an innovative business model, which offers consumers free megabytes of data credited to the data cap, in exchange for engagement with the advertiser by performing some tasks on social media sites, such as viewing social media ads, downloading social media apps, and shopping through social media apps. We provide an analysis of this mechanism as well as useful guidance to policymakers. Chapter 7 discusses some of the key future trends in social media and associated challenges.

This book is written for academicians and professionals involved in social media and social media analytics.

Authors

Subodha Kumar is the Paul R. Anderson Distinguished Chair Professor of Marketing and Supply Chain Management and the Founding Director of the Center for Business Analytics and Disruptive Technologies at Temple University's Fox School of Business. He has secondary appointments in Information Systems and Statistical Science Departments. He also serves as the PhD Concentration Advisor for Operations and Supply Chain Management. He is a board member for many organizations. Prof. Kumar has been awarded a Changjiang Scholars Chair Professorship by China's Ministry of Education. He is also a Visiting Professor at the Indian School of Business (ISB). He has served on the faculty of University of Washington and Texas A&M University. He has been the keynote speaker and track/cluster chairs at leading conferences. Prof. Kumar was elected to become a Production and Operations Management Society (POMS) Fellow in 2019. He has received numerous research and teaching awards. He has published more than 150 papers in reputed journals and refereed conferences. He was ranked #1 worldwide for publishing in Information Systems Research. In addition, he has authored a book, book chapters, Harvard Business School cases, and Ivey Business School cases. He also holds a robotics patent. Prof. Kumar is routinely cited in different media outlets including NBC, CBS, Fox, *Business Week*, and the *New York Post*. He is the Deputy Editor of *Production and Operations Management Journal* and the Founding Executive Editor of *Management and Business Review* (MBR). He also serves on other editorial boards. He was the conference chair for POMS 2018 and DSI 2018, and has cochaired several other conferences.

Liangfei Qiu is the PricewaterhouseCoopers Associate Professor at Warrington College of Business, University of Florida. Prof. Qiu also serves as the PhD coordinator for the Department of Information Systems and Operations Management. His current research focuses on social technology (e.g., social networks, social media, and prediction markets), platform technology (e.g., sharing/gig economy, e-commerce platforms, and healthcare analytics), telecommunications technology, and fintech. Prof. Qiu's research has appeared in premier academic journals and leading media outlets, such as *Bloomberg Businessweek*, *Conversation*, and *ScienceDaily*.

He received the INFORMS Information Systems Society Sandy Slaughter Early Career Award and Association for Information Systems (AIS) Early Career Award in 2019. He is currently an Associate Editor at *MIS Quarterly*, a Senior Editor at *Production and Operations Management*, and an Associate Editor at *Decision Support Systems*.

How Social Media Shapes the Way

1

1.1 INTRODUCTION: "BELLWETHER"

In Connie Willis's famous science-fiction novel, "Bellwether," the main character, Dr. Sandra Foster, studies how to predict fads by experimenting with a flock of sheep. In the novel, the secret to all fads is: "The herd instinct. People wanted to look like everybody else. That was why they bought white bucks and pedal pushers and bikinis" (Willis, 1996, p. 33). Before the age of social media, things did not catch on easily because people interacted at a low level. Even if there was a "bellwether" who could lead the flocks, the size of the flocks that this "bellwether" could reach was rather limited. Social media not only amplifies the herd instinct through intense interactions, but also increases the number of "bellwethers" or social media celebrities.

This book looks at how social media affects the way we think and provides a framework to understand and analyze the impact of social media in various industries using analytics. Recent years have witnessed an unprecedented explosion in social media (e.g., Facebook, Twitter, Instagram, and Pinterest). Social media is not about changing a specific industry, but rather revolutionizing the nature of how businesses operate and changing the landscape of industry competition in online retailing, healthcare, telecommunications, etc.

When one of the authors was writing this book, his cellphone beeped, and the Facebook app reminded him that he checked in at a fancy restaurant in Austin, Texas, eight years ago. It was a valuable memory, especially when restaurant dining became impossible during the COVID-19 pandemic. It also illustrates how social media affects the way we think. Let our social media journey begin!

1.2 HOW SOCIAL MEDIA SHAPES THE WAY COMPANIES AND CONSUMERS THINK

It is well known that aggregating dispersed information from crowds in the right way may produce accurate predictions. The reason is that when we aggregate diverse individuals' estimates, those errors can be canceled out. Surowiecki (2004) called it the wisdom of crowds. However, as people are increasingly influenced by social media, a slight error of an individual tends to be amplified. Therefore, we may observe the madness of crowds instead of the wisdom of crowds. In Chapter 2, we dig deeper into how social media and online social networks affect the operation of the wisdom of crowds.

Since social media affects the way consumers think, it naturally affects how companies think. Social media–based strategies are fundamentally different from traditional firm strategies without considering the impact of social media content. Negative word of mouth on social media stemming from a poor customer engagement or a brand response can disseminate rapidly and reach a large audience. In Chapter 3, we examine the role of social media in firms' strategies.

Social media platforms are prone to abuse and manipulations, manifested in the form of fake news and fake reviews. Even if a statement is known to be false, it becomes less wild if people hear it again and again. Social media platforms spread and repeat fake news faster and farther than ever before. In Chapter 4, we look at social media misinformation and identify who spreads misinformation.

The COVID-19 pandemic caused a significant disruption in the offline healthcare channel, and online health communities illustrate the digital resilience in recovering from and adjusting to this massive exogenous disruption. In recent years, with the rapid development of information technologies, online health communities with social media features are increasingly important in supporting access to the mass of information and resources and promoting barrier-free communication and information exchanges between physicians and patients. Chapter 5 investigates the design of online healthcare communities.

With the upcoming 5G Internet, the Internet of Things (IoT) connects billions of physical devices worldwide, collecting and sharing data for smart cars, smart homes, and smart cities. This fast-growing market provides a huge opportunity, and the telecommunications industry has been exploring new business models. In Chapter 6, we introduce some new business models inspired by social media and telecommunications technology. The last chapter, Chapter 7, discusses some of the key future trends in social media and associated challenges.

REFERENCES

Surowiecki, J. (2004). *The wisdom of crowds: Why the many are smarter than the few and how collective wisdom shapes business, economies, societies, and nations.* New York: Doubleday.

Willis, C. (1996). *Bellwether.* New York: Bantam Books.

Wisdom of Crowds Meets Social Media

2

2.1 INTRODUCTION: GUESSING WEIGHT COMPETITION IN THE AGE OF SOCIAL MEDIA

The year 2021 is the year of the ox in the Chinese zodiac. Interestingly, one of the first stories about the wisdom of crowds is a competition to guess the weight of an ox at a country fair in 1906 (Surowiecki, 2004): Eight hundred people participated in the competition, and no one got a close guess. However, the median (1,197 lb.) of all guesses was extremely close to the actual weight (1,198 lb.) of the ox. The main insight of this story is that some people tend to make positive errors while others tend to make negative errors. When we aggregate diverse individuals' estimates, those errors can be canceled out. In other words, aggregating dispersed information from crowds in the right way may produce accurate predictions.

A critical condition to ensure the wisdom of crowds is that individuals' estimates are independent, which means an individual's guess does not affect others' guesses (Lorenz et al., 2011). However, this condition is rarely satisfied in many real-world scenarios. If individuals' estimates are affected by each other, the herd instinct of human beings may lead to the madness of crowds rather than the wisdom of crowds (Qiu & Whinston, 2017). In the old times, people used to choose restaurants by looking at how many customers were already there. In a study, the ranking information of the five most popular dishes was shown to some customers and the demand for those dishes increased significantly (Cai et al., 2009). The herd instinct of following others also played a great role in the tulip mania, the first recorded speculative bubble: In February 1637, the prices of some

DOI: 10.1201/9781003196198-2

single tulip bulbs were more than ten times the annual income of a skilled artisan (MacKay, 1980). Therefore, if individuals receive more information from others, it may not benefit the crowds' wisdom. Instead, information sharing among individuals tends to corrupt the wisdom of crowds. The reason is that when an individual's guess does affect others' guesses, a small error of this individual will be amplified. The key "error canceling" mechanism mentioned earlier will no longer work (see Figure 2.1).

In fact, individuals' estimates are becoming more and more correlated due to the intense interactions in today's social networking world. Before the age of social media, social interactions were rather limited and were built primarily on face-to-face interactions, emails, and phone calls. Social media and online social networks, such as Facebook, Twitter, Instagram, Reddit, and Clubhouse, have revolutionized how we interact with others. It is widely believed that social media has played an important role in political elections (Aral, 2020; Mallipeddi et al., 2021). Instead of focusing on a small number of friends who are physically close to us, social networking apps allow us to communicate with tenuous ties in this global village.

People are increasingly influenced by social media, and customers turn to social media for product discovery. For example, social commerce has become one of the mainstream retail channels and makes consumers see others' purchase decisions and influence each other more easily. Many social networks have introduced pro-selling features, such as shoppable posts (e.g., Instagram Storefronts), and 72% of Instagram users make purchase decisions after seeing something on others' Instagram pages (Influencer MarketingHub, 2021, Qiu et al., 2021). In the social gallery of Kohl's (a department store), consumers can see posts on Kohl's products from both Twitter and Instagram and purchase them via direct links to the product home page.

Furthermore, location-based social networking applications (e.g., Facebook Places and Yelp) allow consumers to seek their social network friends' recommendations and share their location information (i.e., mobile

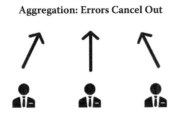

Aggregation: Errors Cancel Out   **Aggregation: Errors May Not Cancel Out**

Wisdom of Crowds without Social Media   **Wisdom of Crowds with Social Media**

FIGURE 2.1   Wisdom of Crowds.

check-ins) with friends through GPS-equipped mobile devices (Qiu et al., 2018). It is time for us to revisit our earlier example: In the old times, people used to choose restaurants by looking at how many customers were already there. In the age of social media, the influence from others is stronger: People can observe their friends' mobile check-ins and know the dining choices made by their Facebook friends without having to visit restaurants to observe their friends' behaviors physically. Researchers show that a consumer's dining decision is significantly affected by the observation of friends' choices in location-based social networks (Qiu et al., 2018).

In many other scenarios, a number of studies present similar findings on peer influence in the age of social media. When readers rate books, their online ratings are significantly affected by the ratings of their friends in online social networks (Wang et al., 2018). A similar phenomenon of peer influence has also been found in online digital music consumption (Hendricks et al., 2012). In these cases, the aggregated opinions, which are represented by the mean ratings, do not necessarily reflect the wisdom of the crowds because any small initial errors can be amplified by peer influence in social media.

Due to the explosive growth of social media, the wisdom of crowds may not always work in the age of social media. An interesting question naturally arises: How do social media and online social networks affect the operation of the wisdom of crowds? In this chapter, we will dig deeper into this issue from several aspects. One challenge in implementing the wisdom of crowds is how to aggregate diverse estimates into a good prediction. A prediction market is one of the most popular information aggregation mechanisms to tap into the wisdom of crowds (Qiu & Kumar, 2017). In prediction markets, the individual participants place bets on events that they think are most likely to happen, thus revealing their private information. Essentially, it is a betting market, which incorporates real-time information and provides a method of "putting your money where your mouth is" (Qiu et al., 2013). Prediction market prices, which are similar to stock prices in financial markets, have informational value because they aggregate the diverse information of market participants and represent overall market forecasts. For example, in the Iowa Electronic Markets, one of the most famous prediction markets, traders can bet whether the next U.S. president is a Democrat or a Republican by buying and selling contracts that pay $1 if a given candidate wins the election (Berg et al., 2008).

There are two types of prediction markets: public prediction markets and corporate prediction markets. A public prediction market is a betting market that is generally available to everyone. Notable examples are the Iowa Electronic Markets and PredictIt, where anyone can join as a trader. A corporate prediction market is an internal betting market for certain employees only. Recently, many big organizations have aimed to take advantage of the small bits and pieces of relevant knowledge dispersed in diverse individuals for better decision-making. Hewlett-Packard, Intel, BestBuy, Microsoft, Ford, Chrysler, Google, Eli Lilly, General Electric, and Siemens have all employed internal prediction markets (Qiu et al., 2017). For example, Hewlett-Packard uses corporate prediction markets to forecast sales, and the predictions consistently beat the official forecasts (Chen & Plott, 2002). At Siemens, corporate prediction markets are used to forecast project completion dates (Leigh & Wolfers, 2007). Intel uses corporate prediction markets to predict sales as well as the allocation of manufacturing capacity (Gillen et al., 2017). At Ford, corporate prediction markets are used to forecast sales volumes, product features, and release timing (Cowgill & Zitzewitz, 2015).

The rest of this chapter is organized as follows. In Section 2.2, we will examine how social image (what other people think of the individual) in social media influences the wisdom of crowds in public prediction markets. In social media, individuals desire to be liked and respected by others, which provides a strong motivation for them to guess correctly in the competition. In Section 2.3, we look at how peer influence in social media impacts the wisdom of crowds in the context of corporate prediction markets. In fact, sometimes, peer influence in social media may benefit the wisdom of crowds in this context. In Section 2.4, we investigate the free-rider effect when people utilize the wisdom of crowds in a social network setting. If individuals can receive useful information from their friends, they might be less likely to acquire information by themselves when they form their guesses. More importantly, if information acquisition is costly, information sharing in a social network need not undermine the wisdom of crowd effect.

2.2 PUBLIC PREDICTION MARKETS AND SOCIAL MEDIA

We start with examining the impact of social media on public prediction markets. More broadly, a public prediction market can be treated as a user-generated content platform. One of the key tasks in such a platform is to incentivize users to provide high-quality user-generated content and improve individual prediction accuracy. One classical way to incentivize high-quality predictions from individuals is to use monetary incentives like betting on horse racing. However, because online gambling is outlawed in the United States through federal laws and many state laws, most public prediction markets operate with "play money" instead of "real money." Therefore, classical monetary incentives cannot be directly applied to the context of public prediction markets, and participation in public prediction markets mainly relies on voluntary contribution.

In this context, nonpecuniary incentive design is critical in the design of public prediction markets. In particular, in a social media–based prediction market, we can use social image as a nonmonetary incentive to improve individual prediction accuracy. Typically, public prediction market designers have less authority and control over participants than the managers of corporate prediction markets have. Hence, it is particularly helpful to use social image to incentivize public prediction market participants to provide accurate predictions. In many different contexts, researchers have shown that social image can incentivize the provision of user-generated content (Pu et al., 2020; Pu et al. 2021; Shi et al. 2021). For example, Amazon has given users the option to share their identity when posting reviews, while Yelp and TripAdvisor have integrated user accounts with their Facebook accounts. Furthermore, social image concerns can significantly affect individual contributions on GitHub, one of the most prominent open-source software communities (Moqri et al., 2018). More generally, neuroscientists find that gaining social status and social image creates similar brain network responses to monetary stimuli (Payne, 2017).

To empirically examine the role of social image, we look at a social media–based prediction market, where the predictions made by individual participants are pushed to their followers' Twitter timelines (Qiu & Kumar, 2017). In this public prediction market, participants can log onto the platform via Twitter and make predictions by clicking on either the "YEA" (the stated event is likely to occur) or "NAY" (the stated event is unlikely to occur) button. The participants' predictions are automatically tweeted as tweets and pushed to their followers' timelines. When the event being predicted is realized, a second tweet (either "You were right!" or "Your prediction was wrong!") is automatically posted and pushed to the followers' timelines.

In this social media–based prediction market, Twitter followers know whether or not the participant has predicted events correctly, which could incentivize the participant to make predictions more carefully to maintain their reputation. Qiu and Kumar (2017) estimate the effect of audience size (i.e., the number of Twitter followers) and online endorsement (i.e., retweets and likes on Twitter) on users' prediction accuracy. A startling discovery is that an increase in the number of Twitter followers of a user leads to an improvement in individual prediction accuracy in the social media–based prediction market. In addition, an increase in the level of online endorsement also leads to an increase in individual prediction accuracy.

These interesting results illustrate the role of social image in the wisdom of crowds. Without any monetary incentives, merely having the predictions broadcast to more followers leads people to make their predictions more carefully. The number of followers or social endorsement does not add any new information, but it changes the social incentives of prediction market participants to contribute more time and effort in acquiring private information and in improving individual prediction accuracy. In other words, increasing audience size and social endorsements enhances the social image value of broadcasting correct predictions to Twitter followers and, as a consequence, leads to more efforts in making predictions, which further contribute to a higher level of individual prediction performance.

Social image refers to an individual's tendency to be motivated by others' perceptions. In a social media–based prediction market, participants can receive social approval of their behavior by broadcasting correct predictions and gaining admiration that is expressed by others. In contrast, broadcasting wrong predictions to others can lead to social disapproval, which causes embarrassment and shame. Since social image reflects what other people think of the individual, it depends critically on visibility (audience size). If providing an accurate prediction gives social rewards to a participant, a larger audience size will increase the social rewards of doing so. In this social context, prediction market participants want to be perceived as "masters of predictions" and be respected by their Twitter followers. The social image motivation incentivizes those participants to contribute more efforts in making predictions. Furthermore, social image incentives are more important when the density of social interaction among people is greater. Therefore, like audience size, social endorsements also enhance social rewards of proving accurate predictions and lead to more efforts in making predictions.

An empirical challenge of estimating the causal effect of audience size on users' prediction accuracy is that the audience size is endogenously affected by many unobserved confounders, such as personal abilities. Prediction market participants tend to respect and become friends with those with high predictive ability. Establishing a credible casual relation is important in the design of prediction markets. In a typical archival data set, observations are generated by a process such as "users who make more precise predictions also interact with other online users frequently and have a large audience size," but Qiu and Kumar (2017) investigate what happens if the data generating process is changed, such as an increase in the frequency of social interaction level and the audience size. They measure the causal effect of such a treatment.

To address the causality issues, Qiu and Kumar (2017) exogenously manipulated the audience size and/or the level of online endorsements. In particular, they randomly selected some users from the participants of the social media–based prediction market. Among these users, half of them were assigned to the control group, one-fourth of them were assigned to the treatment group with the audience size effect, and one-fourth of them were assigned to the treatment group with the audience size and online

Control Group:
50% of participants, not exposed to any treatment

Treatment Group 1:
25% of participants, audience size treatment

Treatment Group 2:
25% of participants,
audience size and online endorsement treatments

FIGURE 2.2 Experimental Design.

endorsement effect (see Figure 2.2). To generate an audience size effect, they managed some synthetic Twitter accounts and gradually made each of these synthetic Twitter accounts follow the participants in both treatment groups. In addition, for participants in the treatment group with the audience size and online endorsement effect, they used the synthetic Twitter accounts created earlier to like and retweet the treated participants' tweets. The control group was not exposed to any treatment.

Qiu and Kumar (2017) adopted a difference-in-differences (DID) method to empirically estimate the impact of audience size and online endorsement on individual prediction accuracy. The DID method controlled for unobserved but fixed individual characteristics. Their regression model was as follows:

$$PA_{it} = b_0 + b_1 AS_i + b_2 ASOE_i + b_3 Post_t + b_4 (Post_t \times AS_i)$$
$$+ b_5 (Post_t \times ASOE_i) + e_{it}, \tag{2.1}$$

where the dependent variable, PA_{it}, is participant i's prediction accuracy at time t ($t = 0$: pre-treatment period; $t = 1$: post-treatment period), AS_i is a binary variable indicating whether a participant is assigned to the treatment group with the audience size effect, $ASOE_i$ is a binary variable indicating whether a participant is assigned to the treatment group with the audience size and online endorsement effect, $Post_t$ is a binary variable indicating whether it is a post-treatment period.

The coefficients b_1 and b_2 measure the baseline differences between the treatment and control groups before the treatment. The coefficient b_3 captures the time trend common to both treatment and control groups. The coefficients b_4 and b_5 represent the causal effect of the treatments AS_i and $ASOE_i$. The DID method relies on the "parallel paths" assumption, which assumes that the average change in the control group represents the counterfactual change in the treatment group had there been no treatments. The empirical design of Qiu and Kumar (2017) ensures that the "parallel paths" assumption holds. First, the prediction market participants in the control and treatment groups were randomly selected. Second, the treatments in audience size and online endorsement were exogenous shocks because the treated users were followed by synthetic Twitter accounts. After estimating the regression model (2.1), they found that $b_4 = 15.5\%$ (p-value < 0.05) and $b_5 = 19.6\%$ (p-value < 0.01), indicating that more Twitter followers increase the individual prediction accuracy by 15.5%, and the extra online social endorsement leads to an additional 4.1% (19.6% – 15.5% = 4.1%) increase in individual prediction accuracy. These results highlight the role of social image in incentivizing participants to make precise predictions: To gain admiration that is expressed by others, prediction market participants tend to spend more time and effort to make careful predictions.

To explore the heterogeneous treatment effects, Qiu and Kumar (2017) also split their sample into subsamples according to different levels of initial indegrees and outdegrees of prediction market participants and rerun regression (2.1). They find that the treatment effect of audience size on individual prediction accuracy is much stronger for the top 50 percentile outdegree group than for the bottom 50 percentile outdegree group. A possible reason is that the treatment of audience size incentivizes prediction market participants to use their information-sourcing channels (outdegree) to improve the prediction accuracy. Participants who belong to the top 50 percentile group have a larger number of information-sourcing channels (outdegree), and hence they can more effectively improve their prediction accuracy.

In this randomized field experiment, Qiu and Kumar (2017) establish the causal effect of audience size and online endorsement on prediction market participants' prediction accuracy. In the prediction market practice, a designer can use the archival data to estimate the correlation between prediction accuracy and audience size. However, a causal relationship is more useful than a pure correlation when we want to provide guidance for a prediction market designer about whether to incorporate social media into their original prediction markets. In reality, individuals tend to become friends with participants with high predictive ability, which may overestimate the social effect. They use an exogenous variation in the audience size and social endorsements to identify the causal impact and help prediction market designers quantify the benefit of incorporating social media into prediction markets.

An important implication of the results of Qiu and Kumar (2017) is that the integration of social media into prediction market systems may effectively incentivize participants to achieve better forecasting performance without providing monetary incentives. Previous studies show that monetary incentives encourage prediction market participants to search for the best information (Chen & Plott, 2002). Surprisingly, social image from social media can also create effective incentives for prediction market participants to make accurate predictions.

For practitioners looking to implement their own prediction markets, the results of Qiu and Kumar (2017) illustrate the usefulness of introducing social technologies and integrate social media into prediction markets. Their empirical evidence supports that social image plays a crucial role in improving participants' prediction accuracy and incentivizing participants to supply thoughtful responses. Besides prediction markets, their results also have broader implications on incorporating social technologies into the design of online community-based question answering sites, such as Quora and Stack Overflow. Social image from online communities may also motivate users to contribute high-quality content in these sites.

2.3 CORPORATE PREDICTION MARKETS AND SOCIAL MEDIA

A persistent challenge for large companies is to find a way to take advantage of the knowledge dispersed in various parts of their organizations. The traditional decision-making process in companies is generally hierarchical. The manager of each business department is responsible for acquiring all information from that department, synthesizing it, and making decisions or reporting it up the chain of command. However, the corporate hierarchy may distort the useful information dispersed in the organization. In particular, although some low-ranking employees may have valuable information to share, they may conform to official reports or the opinion of high-ranking managers, which is known as the "hidden profiles" effect (Stasser & Titus, 2003). When employees report their opinions to upper-level decision-makers in the corporate hierarchy, they may become "yes men," who tend to conform to the opinion of their supervisors and place a larger weight on existing public information than justified by its informational content (Qiu et al., 2017).

A corporate prediction market is an alternative way to facilitate the flow of information around the hierarchy by exploiting the wisdom of crowds within the company. In a corporate prediction market, a risky asset reflects an issue of interest to the company, such as sales trends of a new product or the product's readiness to launch. As a collaborative market mechanism, a corporate prediction market augments the traditional hierarchical decision-making process and improves overall decision quality. In other words, corporate prediction markets can help reduce the costs of hierarchy by allowing information to flow to top decision-makers more efficiently (Abramowicz & Henderson, 2007). As discussed in Section 2.1, a number of big companies are experimenting with corporate prediction markets, and early evidence on the performance of corporate prediction markets has been encouraging: In many cases, corporate prediction markets outperform existing forecasts in terms of prediction precision (Gillen et al., 2017).

At first glance, a corporate prediction market might help alleviate the hidden profile effect: the overreaction to existing public information. In a typical corporate prediction market, participants can trade anonymously and have incentives to correct the conformity and overreaction to public information. Therefore, the anonymous trading system in prediction markets can make managers discover employees' uncensored opinions. In a corporate prediction market, low-ranking employees who dissent from an official report or the opinion of a high-ranking manager would have financial incentives to trade against it. Like trading in financial markets, this anonymity approach may potentially give a voice to employees who otherwise would be unwilling to speak out due to the effect of hidden profiles.

However, more recent research shows that a corporate prediction market cannot completely solve the problem of overreaction to existing public information (Qiu et al., 2017). There are two types of information within an organization. The first type is public information, such as official company reports and opinions of high-ranking managers, which is known to all internal employees. In Google's prediction markets, the online summaries of project status are typically visible to all Google employees and can be treated as public information (Coles et al., 2007). The second type is private information, which is acquired only by individual employees, such as tacit knowledge from their working experience. A manager on the search quality team at Google mentioned that when a prediction market was related to her own projects, she had private information (Coles et al., 2007).

As mentioned earlier, the anonymity of corporate prediction markets can correct the effect of hidden profiles at the individual level and incentivize employees to express their uncensored opinions through the market trading mechanism. However, the information-aggregation mechanism in a corporate prediction market tends to place a larger than efficient weight on public information and leads to another type of hidden profiles effect at the aggregate level. The reason is that the presence of public information might distort the formation of prediction market prices. In a prediction market, all participants receive the same public information, which conveys useful information on the uncertain underlying event. All participants tend to place a certain weight on public information as their best guesses. Then, the market mechanism aggregates all participants' guesses. Therefore, public information is counted multiple times, and the information-aggregation mechanism places a larger than efficient weight on public information. Any noise contained in the public information tends to be amplified by overweighting public information.

How to mitigate the problem of overweighting public information in a corporate prediction market? Qiu et al. (2017) find that social interactions among prediction market participants help correct the overreaction to public information and improve the prediction market performance. As mentioned earlier, the problem of overweighting public information is caused by the fact that the information aggregation process counts public information multiple times and magnifies the noise contained in public information. Social interactions among participants facilitate private information exchange among participants, which leads to a larger weight on private information and a smaller weight on public information. Therefore, to some degree, social interactions among prediction market participants help correct the problem of overweighting public information.

This is particularly relevant for corporate prediction markets in the age of social media. Unlike a public prediction market, participants in a corporate prediction market are internal employees, and they are socially connected through private discussions in the office and social media platforms (Qiu et al., 2014a, b). Due to the explosive growth of social technologies, employees are becoming more active on public social media platforms, such as Twitter, Facebook, and LinkedIn, for work-related purposes (Parise et al., 2015). Internal social media platforms are developed by a number of companies, such as 7-Eleven, Capital One, and Dow Chemical, to promote information exchange and social relations among employees (Mello, 2014). Therefore, in the age of social media, a corporate prediction market is likely to be a socially embedded market in which participants frequently share information with their social connections. The advancement of social technologies can facilitate information communications within organizations and alleviate the problem of overweighting public information.

To conclude, the wisdom of crowds hypothesis assumes that prediction market prices can efficiently aggregate diverse information of individuals. However, this hypothesis primarily focuses on the aggregation of private information without considering the role of public information common to all. In fact, public information is a double-edged sword. On the one hand, it conveys useful information about the predicted event. On the other hand, the noisiness of public information is magnified in prediction markets due to the overreaction to public information. The integration of corporate prediction markets with social media helps correct such inefficiency and improve prediction market performance. In particular, corporate prediction market designers can promote social networking among employees using Facebook, Twitter, LinkedIn, or the in-house corporate social media platforms.

2.4 FREE-RIDER EFFECT AND SOCIAL TECHNOLOGIES IN PREDICTION MARKETS

In an increasingly interconnected world, Twitter-embedded prediction markets are proposed to combine the power of prediction markets with the popularity of social media (Qiu et al., 2013). In Section 2.3, we explained that social interactions among prediction market participants help correct the overreaction to public information and improve the prediction market performance. However, if private information acquisition is costly, social interactions among prediction market participants may lead to a side effect: the free-rider effect. If prediction market participants can obtain information freely from their social connections, they are less willing to acquire costly private information by themselves (Qiu et al., 2014a). Suppose all prediction market participants want to free ride on their social connections' effort. In that case, the total input of private information will be much less, which is detrimental to prediction market performance.

Therefore, on the one hand, social interactions among prediction market participants may help alleviate the problem of overweighting public information (a positive effect on prediction market performance). On the other hand, if information acquisition is costly, social interactions can cause the free-rider effect, which may hurt prediction market performance (a negative effect on prediction market performance). A natural question arises: When do social interactions among prediction market participants benefit prediction market performance? In other words, when does the positive effect of social interactions dominate the negative effect?

Qiu et al. (2014a) use a controlled laboratory experiment to examine this question. As mentioned in Section 2.2, an empirical challenge of estimating the impact of social interactions is that social interactions are endogenously determined. In particular, social interactions can be the result of past prediction performance. Researchers have addressed this empirical challenge using different approaches. One approach is to rely on a natural experiment, which is similar to an exogenous variation in real laboratory or field experiments (Zhang & Zhu, 2011). For example, in a study investigating peer influence among college roommates, a natural experiment is an exogenous shock: Freshmen entering a college are randomly assigned to dorms and to roommates. The second approach is to use exogenous instrument variables. For example, the spurious estimates of school-based peer influence can be corrected by instrumenting for peer behavior using the average behavior of the peers' parents (Gaviria & Raphael, 2001). The approach of Qiu et al. (2014a) belongs to the third one. In their controlled laboratory experiment, participants were randomly assigned to different social network positions in prediction markets. The approach of laboratory experiments allowed them to tease out many confounding factors, but it could not fully mirror the prediction market environment in reality. In this sense, the field experiment approach introduced in Section 2.2 can complement laboratory experiments and address the concern of artificial social interactions.

The experimental results of Qiu et al. (2014a) show that social interactions enhance prediction market performance only when the cost of information acquisition is low. They also use simulations to examine bigger and more complex social networks, and the same result holds. As mentioned earlier, there are two underlying mechanisms for social interactions. The first one is that social interactions can alleviate the problem of overweighting public information and enhance prediction market performance. The second one is the free-rider effect, which leads to an inefficient level of information acquisition. When the cost of information acquisition is low, the free-rider problem is not a big concern, and most participants are willing to pay the cost to acquire valuable private information. In this case, the first effect (correcting the overreaction to public information) dominates the second free-rider effect, and overall social interactions can enhance prediction market performance. In contrast, when the cost of information acquisition is high, social interactions impede information acquisition of prediction market participants because of possible free-riding opportunities, thus lowering the forecasting accuracy of prediction markets.

The striking findings of Qiu et al. (2014a) have direct implications for the business practice of prediction markets. When the predicted event is relatively simple, it is typically easier to obtain useful information. In this case, it is beneficial in terms of prediction market performance to encourage social networking and information exchange among participants. However, if the predicted event involves complicated issues, it may not be a good idea to promote social interactions among participants.

In a follow-up study, Qiu et al. (2014b) further investigate how social network structure affects prediction market performance. They differentiate between two types of social network structures. One is a balanced social network in which participants have a similar number of social connections, and the other is a network structure akin to star networks in which central participants have many more connections than peripheral participants (see Figure 2.3). They find that a more balanced social network structure contributes to the success of prediction markets, while network structures akin to star networks are ill-suited to prediction markets.

The results of Qiu et al. (2014b) suggest that information aggregation in prediction markets is inefficient for a star network when some participants are significantly more well connected than most others. Although central participants are in a perfect position (information hub) to aggregate information from other

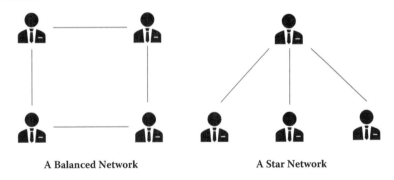

FIGURE 2.3   Network Structure.

peripheral participants, prediction markets fail to aggregate all the guesses efficiently. The wisdom of crowds fails when a small prominent group of opinion leaders exists (Golub & Jackson, 2010). Qiu et al. (2014b) show that a central prediction market participant becomes significantly more influential than peripheral participants, and any noise in a central participant's information tends to be magnified. Furthermore, a central prediction market participant has significantly more connections, which leads to a more serious free-rider problem and reduces total information provision. Such inefficiency is less a concern in a more balanced social network structure.

These experimental results inform prediction market designers to understand social network structures among employees when building corporate prediction markets. Of course, it is not easy to identify employees' underlying social network structure within a large organization. Early work finds that several main factors, such as physical office locations, joint-work relationships, or sharing a common non-English native language, can significantly affect information exchange among employees at Google (Cowgill & Zitzewitz, 2015). Potentially, prediction market designers can dig deep into these proximity measures to infer employees' social network structures within an organization and improve the design of corporate prediction markets for decision support.

REFERENCES

Abramowicz, M. & Henderson, M. T. (2007). Prediction markets for corporate governance. *University of Notre Dame Law Review, 82*(4), 1343–1414.

Aral, S. (2020). The hype machine: How social media disrupts our elections. *Our Economy, and Our Health-and how We Must Adapt*. New York: Currency.

Berg, J., Forsythe, R., Nelson, F., & Rietz, T. (2008). Results from a dozen years of election futures markets research. *Handbook of experimental economics results*, *1*, 742–751.

Cai, H., Chen, Y., & Fang, H. (2009). Observational learning: Evidence from a randomized natural field experiment. *American Economic Review*, *99*(3), 864–882.

Chen, K. Y., & Plott, C. R. (2002). Information aggregation mechanisms: Concept, design and implementation for a sales forecasting problem. California Institute of Technology Social Science Working Paper No. 1131.

Coles, P. A., Lakhani, K. R., & McAfee, A. (2007). *Prediction markets at Google*. Boston, MA: Harvard Business School Case 9-607-088.

Cowgill, B., & Zitzewitz, E. (2015). Corporate prediction markets: Evidence from Google, ford, and firm x. *The Review of Economic Studies*, *82*(4), 1309–1341.

Gaviria, A., & Raphael, S. (2001). School-based peer effects and juvenile behavior. *Review of Economics and Statistics*, *83*(2), 257–268.

Gillen, B. J., Plott, C. R., & Shum, M. (2017). A pari-mutuel-like mechanism for information aggregation: A field test inside Intel. *Journal of Political Economy*, *125*(4), 1075–1099.

Golub, B., & Jackson, M. O. (2010). Naive learning in social networks and the wisdom of crowds. *American Economic Journal: Microeconomics*, *2*(1), 112–149.

Hendricks, K., Sorensen, A., & Wiseman, T. (2012). Observational learning and demand for search goods. *American Economic Journal: Microeconomics*, *4*(1), 1–31.

Leigh, A., & Wolfers, J. (2007). Prediction markets for business and public policy. *Melbourne Review: A Journal of Business and Public Policy*, *3*(1), 7–15.

Lorenz, J., Rauhut, H., Schweitzer, F., & Helbing, D. (2011). How social influence can undermine the wisdom of crowd effect. *Proceedings of the National Academy of Sciences*, *108*(22), 9020–9025.

Influencer MarketingHub. (2021). Social Media Trends for 2021. Last accessed on February 17, 2021: https://influencermarketinghub.com/Social_Media_Trends_for_2021.pdf.

MacKay, C. (1980). *Extraordinary popular delusions and the madness of crowds*. New York: Harmony Books.

Mallipeddi, R. R., Janakiraman, R., Kumar, S., & Gupta, S. (2021). The effects of social media content created by human brands on engagement: Evidence from Indian General Election 2014. *Information Systems Research*, *32*(1), 212–237.

Mello, J. A. (2014). *Strategic human resource management*. Stamford, CT: Cengage Learning.

Moqri, M., Mei, X., Qiu, L., & Bandyopadhyay, S. (2018). Effect of "following" on contributions to open source communities. *Journal of Management Information Systems*, *35*(4), 1188–1217.

Parise, S., Whelan, E., & Todd, S. (2015). How Twitter users can generate better ideas. *MIT Sloan Management Review*, *56*(4), 21.

Payne, K. (2017). *The broken ladder: How inequality affects the way we think, live, and die*. Penguin: New York.

Pu, J., Chen, Y., Qiu, L., & Cheng, H. K. (2020). Does identity disclosure help or hurt user content generation? Social presence, inhibition, and displacement effects. *Information Systems Research*, *31*(2), 297–322.

Pu, J., Liu, Y., Chen, Y., & Cheng, H. K. (2021). What questions are you inclined to answer? Effects of hierarchy in corporate knowledge sharing communities. *Information Systems Research*, Forthcoming.

Qiu, L., Cheng, H. K., & Pu, J. (2017). Hidden profiles in corporate prediction markets: the impact of public information precision and social interactions. *MIS Quarterly*, *41*(4), 1249–1273.

Qiu, L., & Kumar, S. (2017). Understanding voluntary knowledge provision and content contribution through a social-media-based prediction market: A field experiment. *Information Systems Research*, *28*(3), 529–546.

Qiu, L., Rui, H., & Whinston, A. (2013). Social network-embedded prediction markets: The effects of information acquisition and communication on predictions. *Decision Support Systems*, *55*(4), 978–987.

Qiu, L., Rui, H., & Whinston, A. B. (2014a). Effects of social networks on prediction markets: Examination in a controlled experiment. *Journal of Management Information Systems*, *30*(4), 235–268.

Qiu, L., Rui, H., & Whinston, A. B. (2014b). The impact of social network structures on prediction market accuracy in the presence of insider information. *Journal of Management Information Systems*, *31*(1), 145–172.

Qiu, L., Shi, Z., & Whinston, A. B. (2018). Learning from your friends' check-ins: An empirical study of location-based social networks. *Information Systems Research*, *29*(4), 1044–1061.

Qiu, L., Vakharia, A., & Chhikara, A. (2021). Multi-dimensional observational learning in social networks: Theory and experimental evidence. *Information Systems Research*, Forthcoming.

Qiu, L., & Whinston, A. B. (2017). Pricing strategies under behavioral observational learning in social networks. *Production and Operations Management*, *26*(7), 1249–1267.

Shi, C., Hu, P., Fan, W., & Qiu, L. (2021) How learning effects influence knowledge contribution in online Q&A community? A social cognitive perspective, *Decision Support Systems*, *149*, 113610.

Stasser, G., & Titus, W. (2003). Hidden profiles: A brief history. *Psychological Inquiry*, *14*(3-4), 304–313.

Surowiecki, J. (2004) *The wisdom of crowds: Why the many are smarter than the few and how collective wisdom shapes business, economies, societies, and nations*. New York: Doubleday.

Wang, C., Zhang, X., & Hann, I. H. (2018). Socially nudged: A quasi-experimental study of friends' social influence in online product ratings. *Information Systems Research*, *29*(3), 641–655.

Zhang, X. M., & Zhu, F. (2011). Group size and incentives to contribute: A natural experiment at Chinese Wikipedia. *American Economic Review*, *101*(4), 1601–1615.

Social Media and Firm Strategies

3

3.1 INTRODUCTION: REDDIT TRADING FRENZY AND UNITED AIRLINES' BRAND CRISIS

On January 27, 2021, the stock of GameStop, a video game retailer, surged more than 100% and drove GameStop's value up by more than $10 billion (MacMillan & Torbati, 2021). The GameStop trading mania has vividly illustrated the power of social media in the investing community of Reddit, a prominent social news platform. Several traditional hedge funds failed to gather critical information from social media content. In contrast, innovative investment firms have realized the importance of social media in firm strategies and operations. They have started tracking Reddit and Twitter for the sentiment of retail traders. For example, technology firms, such as Quiver Quantitative, Swaggy Stocks, and Robintrack, are using social media to gauge investors' sentiment and detect events quickly. They use machine-learning algorithms and natural language processing methods to analyze social media content, which is sometimes too subtle to detect otherwise, or until it is too late to trade profitably (MacMillan & Torbati, 2021).

Another example of showing the power of social media is the United Airlines' brand crisis. On April 9, 2017, United Airlines faced a brand crisis after a video showed a passenger being removed from one of the airlines' flights by an airport's aviation security officials resulting in physical injuries (Victor & Stevens, 2017). The video went viral on social media, causing both online and offline firestorms. In response to the crisis, the airline issued an

DOI: 10.1201/9781003196198-3

apology statement on April 11, 2017, which was posted to the airline's official Twitter account and website. On the one hand, social media can amplify the profound damages caused by a public relations crisis. On the other hand, the public nature of social media offers an innovative way for firms to handle a brand crisis and manage their customer relationships publicly. Compared to traditional customer relationship management, social media enables firms to directly communicate with their customers, which can strengthen the firm–customer relationships and maintain higher customer engagement levels.

However, social media–based strategies are fundamentally different from traditional firm strategies without considering the impact of social media content. Public engagement with customers on social media does not come without risk. Negative word of mouth on social media stemming from a poor customer engagement or a brand response can disseminate rapidly and reach a large audience, especially when the customer has a strong social influence (Chevalier & Mayzlin, 2006). In Section 3.2, we examine how firms' online management responses in social media websites affect firm performance. Then, we look at how firms generate user engagement in their social media communities and how social media affects firms' pricing and other operations strategies in Section 3.3. In Section 3.4, we discuss a new type of social media, location-based social media, which offers users the option to share the locations with their social media friends.

3.2 ONLINE MANAGEMENT RESPONSES IN SOCIAL MEDIA

Online review platforms, such as Yelp and TripAdvisor, have become in-creasingly social. They provide opportunities for customers to participate in an online community that rates local businesses and service providers. In this type of community, it is crucial for businesses to engage with less-satisfied customers and address their concerns about products or services (Gu & Ye, 2014). Therefore, these platforms have launched new business functions to communicate with customers and respond to online reviews. The business owners' responses are appended to the end of consumers' online reviews, and all consumers can observe these responses publicly (Kumar et al., 2018b). How does this online interaction between business owners and consumers translate into real-world performance?

It is widely believed that online reviews have a significant impact on purchasing decisions of consumers (Kwark et al., 2021; Wang et al., 2021). If consumers' online reviews are important, a natural question is how businesses should react to online reviews. In general, businesses can react to consumers' online reviews in two ways. One is to respond to online reviews, which is the focus of this chapter. The other is review manipulation: Businesses may hire real individuals to post overly positive online reviews to boost demand (Kumar et al., 2018a; Wang et al., 2020), which will be discussed in Chapter 4.

Kumar et al. (2018b) aim to estimate the impact of online management responses on business performance. Intuitively, the introduction of the new online management response feature may benefit those businesses that adopt it. An essential purpose of online management responses is to interact with less satisfied customers and prevent them from experiencing negative emotions. The prevalence of social media is changing the traditional way of engaging customers. It is critical to prevent consumers from experiencing negative emotions because they can easily distribute their negative opinions to large audiences on social media fan pages and create a brand crisis like the case of United Airlines mentioned earlier.

Online management responses may also have spillover effects on nearby businesses. On the one hand, a restaurant's online management responses may attract customers from a nearby restaurant, causing a negative spillover effect. On the other hand, a restaurant's online management responses may help bring more potential customers to the neighborhood and benefit nearby restaurants by generating traffic. Kumar et al. (2018b) make a startling discovery: The spillover effect depends on the competition intensity. If a nearby restaurant and a focal restaurant belong to the same category, for example, both of them are Italian restaurants, then the negative spillover effect dominates: The focal restaurant's online management responses attract consumers from the nearby restaurant. However, if they belong to different categories and are not in direct competition, the positive spillover effect dominates: The focal restaurant's online management responses help bring more consumers to the neighborhood (see Figure 3.1). The reason is that a focal restaurant does

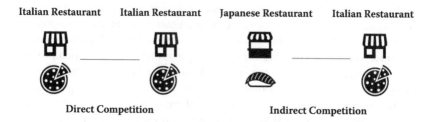

Italian Restaurant   Italian Restaurant   Japanese Restaurant   Italian Restaurant

Direct Competition   Indirect Competition

FIGURE 3.1 Direct Competition and Indirect Competition.

not have intense competition with a nearby restaurant in a different category. In fact, when a focal restaurant attracts more consumers by responding to online consumer reviews, it may unintentionally benefit nearby restaurants.

To estimate the impact of the introduction of online management responses on restaurant performance, Kumar et al. (2018b) estimated the following regression model:

$$
\begin{aligned}
\mathrm{Log}(Checkin_{it}) = {}& c_i + b_0 + b_1 Post_t \\
& + b_2 (Post_t \times Response_{it}) \qquad\qquad + b_3 Controls \\
& + e_{it},
\end{aligned}
\tag{3.1}
$$

where the dependent variable, $\mathrm{Log}(Checkin_{it})$, is the log number of mobile check-ins of restaurant i in month t. In the mobile commerce age, the number of mobile visits is a proxy measure for the popularity of a restaurant (Wang et al., 2015). The variable $Post_t$ indicates whether the online management response feature is introduced, $Response_{it}$ indicates whether restaurant i responds to consumers' online reviews in month t, and c_i is the restaurant-level unobserved fixed effect. Control variables include online review-related information, weather information, and monthly time dummies.

The results of Kumar et al. (2018b) show that $b_1 = -0.0128$ and $b_2 = 0.0972$, and both coefficients are statistically significant. It suggests that after the introduction of the online management response feature, the restaurants that actually respond to online consumer reviews have an 8.44% (9.72% − 1.28%) higher level of performance. In contrast, the restaurants that can respond to consumer reviews but choose not to do so have a 1.28% lower performance level. These results imply that, in general, the introduction of the online management response feature can improve business performance. However, it does not benefit all restaurants. It benefits the restaurants that actively respond to online consumer reviews, but it hurts the restaurants that choose not to do so.

A key empirical challenge in this archival dataset is to establish the causal effect of online management responses on performance. The restaurants that respond to online reviews could be systematically different from the restaurants that do not. For example, a restaurant may choose to respond to online consumer reviews due to low online review ratings or high competition pressure. Without accounting for the endogenous decision to respond to online consumer reviews, one may overestimate the benefit of online management responses. More importantly, actively monitoring and responding to online consumer reviews is a resource-intensive and time-consuming endeavor, especially for small local businesses. It is unclear whether local small businesses should embrace it if the benefit is not significant. Therefore,

Kumar et al. (2018b) address this concern using various empirical matching strategies.

First, Kumar et al. (2018b) combine their regression model with the propensity score matching method. For each treated restaurant that responds to online consumer reviews, they match it with a similar control restaurant that does not respond to online consumer reviews in terms of observable characteristics, such as zip-code level information, competition intensity, and restaurant categories. They run a logit regression based on these characteristics and obtain a predicted propensity score for each restaurant. Next, they match each treated restaurant with the control restaurant with the closest propensity score. Based on this matched sample, they re-estimate regression equation (3.1) and find that their results are robust. The matching procedure ensures that the control restaurants that do not respond to online consumer reviews are more comparable to the treated restaurants that respond to online consumer reviews.

Second, Kumar et al. (2018b) make use of a quasi-experiment to help establish the causality and address unobserved confounding factors. The propensity score matching method can effectively handle observable confounding factors, but what about unobserved confounders? For example, restaurants that actively respond to online reviews may also be more likely to use advertisements to reach audiences. These unobserved restaurant promotions and advertisements are time-varying and may change the differences between the control and treatment restaurants over time. As mentioned in Chapter 2, the gold standard of establishing causality is to run randomized field or laboratory experiments. However, randomized trials are not always available in the business world. In the context of Kumar et al. (2018b), it is hard to imagine that they could randomly assign restaurants to respond to online consumer reviews or not. Therefore, they adopted a quasi-experiment research design.

The intuition of the quasi-experiment design of Kumar et al. (2018b) was to obtain a more similar control group in terms of unobserved time-varying confounders by looking at the time sequence of online management responses. Restaurants that respond to online consumer reviews might be very different from those that do not respond to online consumer reviews in various unobserved characteristics. A better control restaurant is a restaurant that has not responded to online reviews but will do so in the future. Based on this idea, Kumar et al. (2018b) used the looking ahead-propensity score matching method (Bapna et al., 2018) to create a more balanced matched sample. Then, they re-estimated regression equation (3.1), and found that their results were consistent.

The results of Kumar et al. (2018b) enhance our understanding of business owner engagement with customers using social media. The benefit of online management responses is not observed in a consistent manner across all businesses. Only restaurants that choose to respond to online consumer reviews observe increases in business performance. Ignoring the opportunity to engage with customers on social media is not bliss for businesses. Small local businesses may consider investing resources into engaging with consumers through social media by recruiting personnel or hiring third-party, public relations companies that can handle social media and respond to online consumer reviews. Furthermore, The spillover effect of online management responses suggests that local businesses may want to be aware of how their competitors are engaging with consumers online to optimize their own digital strategies. Social media platforms could also help by developing new features and analytical algorithms for businesses to follow online management responses of nearby businesses.

3.3 FIRM STRATEGIES AND SOCIAL MEDIA PLATFORMS

The popularity of social media has fundamentally shaped companies' strategies in different aspects. We first look at how social media helps companies build customer loyalty and generate user engagement. Next, we examine how social media affects firms' pricing and other operations strategies. Finally, we investigate the impact of social media on transactions in marketplaces.

3.3.1 Social Media and User Engagement

Many well-known brands have active online social media communities, but how can new firms starting from scratch attract social media fans and generate user engagement? It is a challenge for new firms to know which types of social media posts initiated by firms are likely to generate the most engagement. Therefore, Bapna et al. (2019) looked at how social media strategies can increase the engagement level of online communities. They collected 15 young retailers' 9,470 Facebook posts and analyzed the relationship between post content and social media engagement (i.e., Facebook "likes") in response to the posts. A unique advantage in their analysis is that they used Amazon Mechanical Turk workers to examine the post content manually instead of adopting automated natural language processing. This procedure allowed them to catch nuances in videos and images that automated algorithms would have missed because humans can easily understand the meaning of video and image content.

Bapna et al. (2019) find that firms' social media strategies should be different for niche products and mass appeal products. Facebook posts about firm milestones, partnerships, or awards generate the most social media engagement for firms that sell niche products. In contrast, firms that sell mass appeal benefit more in terms of engagement from posts that convey industry knowledge, with news, anecdotes, or feature articles about the industry. These results provide nuanced insights into post content and guide firms to design their social media posting strategy effectively.

Besides firms' posting strategies, Kumar et al. (2017) investigate another emerging social media strategy: trademarking Twitter hashtags. For businesses, one tool to engage with their consumers on social media is the use of social media hashtags. Hashtags are used on numerous social media platforms and serve as an indicator for groups of posts related to a particular topic. Firms use social media hashtags as ubiquitous conversation starters, marketing campaign tools, and brand symbols. Creating an original hashtag can give a firm control over a specific social media space in which it can show off its industry expertise, drive viral conversations, and promote a special event, contest, or campaign to interested consumers. For example, the #SmileWithACoke campaign on Twitter used the hashtag to encourage people to share images of special products with names printed on them (Narang, 2017).

However, hashtags have also created challenges for brands because these social media campaigns can be "hijacked" by competitors, distorting a business's original intent or diluting the intended official message. One such example is that a former designer of Fraternity Collection (FC) LLC started using hashtags (#FratCollection and #FraternityCollection) for promoting its own product t-shirts on Instagram. The hashtags used by the designer were similar enough to the brand name of Fraternity Collection (FC) LLC. The United States Patent and Trademark Office (USPTO) confirmed the registration of hashtags as trademarks in 2013. If a trademarked hashtag is misused by a competitor, the owners can file a lawsuit against the competitor for trademark infringement or trademark dilution. An increasing number of firms have been trademarking hashtags and taking advantage of this extra protection (Chu, 2017).

The purpose of firm-generated hashtags is to engage consumers and cultivate an online brand community. However, without trademarking hashtags, firm-generated hashtags may not function properly. As mentioned earlier, hashtags might be "hijacked" by competitors, and social media platforms broadcast such interactions to the whole online brand community, which could severely damage the credibility of firms. Kumar et al. (2017)

empirically examined the impact of trademarking a hashtag on a firm's social media audience engagement and found that trademarking hashtags plays a pivotal role in increasing social media audience engagement in terms of the number of retweets. This result implies that trademarking hashtags helps firms depict themselves as credible by displaying capability and commitment.

Furthermore, Kumar et al. (2017) show that the positive effect of trademarking hashtags on social media audience engagement is stronger for less-known firms with fewer Twitter followers. A possible reason is that consumers may not be familiar with small firms and their brands. Without trademark protection, other competitors can easily use similar hashtags to mislead consumers. In contrast, large firms have well-established brands, and it is more difficult to mislead consumers even in the absence of trademark protection. This finding highlights the heterogeneous effect of trademarking hashtags. For small and less popular firms, trademarking hashtags is an effective and attractive social media control strategy to increase firms' social media audience engagement.

Besides Facebook posts and tweets we have already discussed, another important format of social media is online videos shared on YouTube and other social media sites. Online videos have changed the landscape of marketing and entertainment. YouTube is not only a social media video site, but also a platform where people can launch their careers and become social media celebrities. Qiu et al. (2015) analyzed how social learning and network effects drive the diffusion of online videos on YouTube. Social learning means that viewers receive various information from friends, which can help them infer video quality. The fact your friends have watched a video tells you it is likely to be good. A network effect is a very different mechanism. Whether a video has a good quality or not, people may have strong incentives to watch it if others are watching it. It may become a topic of conversation that can be discussed in social encounters. Many YouTube videos go viral with only pointless-seeming jokes, funny pictures, weird scenarios, or even offensive pranks, but they bring people together around something.

Interestingly, Qiu et al. (2015) find that the two mechanisms mentioned above work for different types of videos. In reality, we see two types of viral videos. One group consists of high-quality videos that feature engaging scenes and articulated storylines. The other group of attention-grabbing videos typically include controversial elements, which provoke controversy and stir heated discussion. Qiu et al. (2015) demonstrate that the popularity of high-quality videos mainly depends on social learning, while network effects drive low-quality but attention-grabbing videos to go viral. This finding is consistent with the results in prior experimental studies: Content that evokes high-arousal emotions (e.g., awe, anger, and anxiety) is more viral (Berger & Milkman, 2012). A direct implication of this result is about managing the influence of consumer buzz in social media. Social media platforms, such as YouTube, can play a much greater role in encouraging the creation of original content by leveraging the multiplier effect of both social learning and network effects. These findings can also help content providers craft contagious content and produce viral videos.

3.3.2 Social Media and Pricing Strategies

Now, we examine how social media affects firms' pricing strategies. The increasing pervasiveness of social networks allows customers to share purchase behaviors with their social network friends. Socially shared purchases have become a mainstream activity of customers and one of the top drivers for online sales (Li et al., 2020; Qiu & Whinston, 2017). In the age of social media, people are connected by a social network, and they can observe their friends' purchase behavior, which may reveal helpful information about the products purchased by their friends. This phenomenon is not new, but undoubtedly, social media can amplify this type of learning behavior.

For example, in the pre-social media era, people used to decide whether to dine based on how many consumers were already in a restaurant. However, nowadays, people can observe their social network friends' mobile check-ins (at restaurants) on Facebook, WeChat, and Yelp. Naturally, consumers learn useful knowledge of restaurants by observing their friends' choices, which is termed as observational learning in the literature (Banerjee, 1992). How does this type of observational learning affect sellers' pricing strategies? In particular, When designing its optimal pricing policy, a seller could control the intensity of observational learning using different pricing strategies. More specifically, a seller faces two trade-offs of choosing different pricing strategies: (i) A monopolistic seller faces a downward-sloping demand curve. Setting a higher price leads to a lower level of quantity demanded, which is the conventional static trade-off discussed in the pricing literature and (ii) A dynamic trade-off involves consumers' learning. If the initial price is too high, then very few early consumers will adopt the product, and the monopolistic seller will not be able to use observational learning to boost its sales in the later periods. On the other hand, if the price is too low, the effect of observational learning is also limited because late consumers would know that their friends purchased the product because of the low price instead of its high quality.

Therefore, the intensity of observational learning is endogenously determined by different pricing strategies set by sellers. Qiu and Whinston (2017) showed that introductory discounts could prevent observational learning, and hence are not always an effective method to boost purchases, which is a striking result. On the other hand, sellers can benefit from the informative pricing strategy that results in more observational learning in social media. The reason is that introductory discounts could be detrimental to social contagion: If a seller charges such a low price initially that almost all consumers adopt the product, this price would not reveal any quality information of the product to future consumers. In this case, observational learning will play no role in increasing the future consumers' willingness to pay.

Recently, online retailers are experimenting with new pricing and other operations strategies (Cheng, 2020; Ding et al., 2021). One of them is social pricing: Online retailers encourage consumers to leverage their social network to get better discounts. For example, the leading e-commerce platform in China, Pinduoduo, provides discounts to consumers during the checkout stage if they invite friends from their social networks to "bargain" for their purchase. The invitees only need to click the "bargain" button (rather than purchasing the product as in group buying) to activate the discount for the consumers. Gao et al. (2020) conducted randomized field experiments to understand the value and mechanisms of social pricing. They find that compared with regular firm-offered discounts, social pricing can significantly increase purchase frequency and order value per purchase of existing consumers. In addition, they reveal that reciprocity is the primary driver of consumer social interactions in social pricing. It may seem intuitive that price discounts can increase sales. However, a striking thing about their findings is that social pricing is more effective than regular retailer-initiated discounts, highlighting the power of this innovative social pricing.

3.3.3 Social Media and Transactions in Marketplaces

The growing importance of social media provides fertile ground for us to understand fundamental constructs of human behavior, such as trust and forgiveness, from transactions in marketplaces. In this chapter, we focus on two questions: (i) Are people more willing to forgive defections in market transactions when they are social media friends? (ii) Does expressing forgiveness discourage future offenses in the presence of social ties?

In terms of the first question, strong social ties on social media can serve to repair and enhance relationships, and people may thus be more willing to forgive their social media friends' defections. However, it is also likely that people are less likely to forgive because they have positive expectations based on prior experiences with that social media friend. When those expectations are not met, lower levels of forgiveness will ensue. Both cases are theoretically plausible, which presents a viable opportunity for our empirical tests.

Regarding the second question, expressions of forgiveness often convey a willingness to maintain or build positive relationships with offenders. In contrast, offenders who have not received forgiveness may view their relationship as irreparably damaged and may, therefore, have little incentive to treat the victim better in the future. However, in repeated interactions, forgiveness may also encourage future defections, and expressing forgiveness could cause problems for victims if offenders get the false impression that they will bear no consequences for their defections.

An empirical challenge in answering these two questions on forgiveness in the context of social media is the task of accounting for endogenous social ties. The reality, and possible confound, is that repeated interactions may create a context in which social ties can emerge as the outcome of repeated relationships. Therefore, empirical evidence has been limited in part because it is hard to generate ideal field data that permits researchers to separate whether successful long-term relations between trading parties are driven either by reputation built through repeated play or by social ties that emerge endogenously from long-term interactions.

To address the challenge of endogenous social ties, Bapna et al. (2017) designed a field experiment that used data from Facebook to measure social ties between subjects, and subsequently, deployed a randomized experiment. Specifically, they manipulated the nature of connectivity between individuals such that some pairings were between socially connected friends on Facebook, while others played anonymously and were only connected through repeated transactions. They then tested how the manipulation of connectivity impacted observed forgiveness outcomes.

Bapna et al. (2017) also conducted a randomized field experiment on Facebook using the investment game, a well-validated game that provides economic measures of trust and forgiveness (Berg et al., 1995). By conducting this experiment within the Facebook social network, they gained access to quantifiable social measures, which allowed them to categorize different levels of social ties between participants. Further, by recruiting participants along with their friends on Facebook, rather than running the game within artificially contrived networks, they increased the external validity of their results.

The experimental results of Bapna et al. (2017) reveal the startling answers to the two questions. First, people are more willing to forgive their friends' defections than forgive anonymous participants' defections. In other words, the level of forgiveness is higher in the presence of social ties. Second, when individuals are anonymous, forgiveness tends to encourage future defections; however, forgiveness can deter future offenses in the presence of social ties. These results on forgiveness suggest that introducing social interactions in the design of online reputation mechanisms can make the cooperative equilibrium more stable. Unexpected shocks (weather uncertainty, supply chain disruptions, etc.) may cause defections from the e-vendor side. The forgiveness needed to make economic exchange possible and stable comes mainly from digital social ties in electronic commerce.

E-vendors have two strategies to enhance their ability to build social capital digitally and to make use of forgiveness in the presence of social ties. First, e-vendors can encourage social interactions in their online forums. For example, they can engage with less satisfied customers, respond to their online comments, and work to influence their future experience and satisfaction positively. Second, e-vendors can spend time and resources embracing new online engagement features supported by social media platforms such as Facebook, WhatsApp, and WeChat. For instance, they may use Facebook social ties as the basis for trusted exchange. These social ties in social media platforms are digitized representations of our physical-world social capital.

3.4 LOCATION-BASED SOCIAL MEDIA

Recently, mobile applications have offered users the option to share their location information with friends (Ghose, 2018). As mentioned earlier, people can check in at restaurants using a mobile application and post their mobile check-ins on their social network accounts. Therefore, consumers can infer the quality of restaurants by looking at their social network friends' mobile check-ins on Facebook, WeChat, and Yelp. However, how do we empirically quantify the impact of observational learning in the real world? Essentially, from data, we observe the correlated behavior among friends. For instance, my friends and I visit the same restaurant. This pattern of correlated behavior among friends can be explained by a number of factors, and observational learning is an informational explanation of causal social influence: An individual's decision is affected by the observation of friends' choices because of their informational content.

Estimating the effect of observational learning is complicated by several plausible confounders. The most important confounding mechanism is the homophily-driven diffusion process: Similarities in friends' tastes can cause correlated friends' choices (Aral et al., 2009). Unlike observational learning, homophily is a non-causal mechanism. Under homophily, correlated friends'

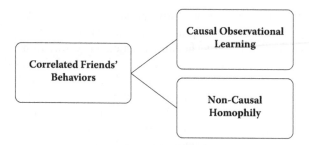

FIGURE 3.2 Mechanisms of Correlated Friends' Behaviors.

choices are driven by similar tastes and characteristics among friends rather than causal social influence among friends (see Figure 3.2). Separating causal social influence (e.g., observational learning) from non-causal mechanisms is essential to the formulation of effective social contagion management policies. The success of viral marketing strategies depends on whether (and when) friends causally influence one another. Without separating causal social influence from non-causal mechanisms, we may overestimate the value of social contagion management policies and network-based marketing.

How can we separate observational learning from other non-causal mechanisms? Suppose we observe a sharp decline in the clustering of mobile check-ins among peers as consumers proceed from trial to repeat. In that case, it will be consistent with a significant observational learning effect in trials because personal dining experience (experienced information) substitutes for observational learning (observed information) from peers (Qiu et al., 2018). In other words, the separation of observational learning from homophily depends on the static nature of the latent homophily effect versus the dynamic nature of the observational learning effect. Homophily is the intrinsic preference similarity between friends based on static (time-invariant) personal characteristics and tastes. Therefore, the effect of correlated tastes (homophily) should be relatively stable over time. In contrast, if the impact of friends' check-ins is mainly driven by the effect of observational learning, we would expect to observe that after a consumer has visited a restaurant, the consumer should rely less on friends' check-ins to infer the restaurant quality. Essentially, when a consumer has a better knowledge of the restaurant (personal dining experience), observational learning becomes less critical.

Using the logic above and data from a location-based application, Qiu et al. (2018) separated observational learning from homophily. Their empirical results suggest that the effect of observational learning drops sharply as consumers proceed from trial to repeat. It is more valuable for newer local vendors to reward their customers' social check-ins, and the observational learning effect is strong in the early introductory months. Additionally, if local vendors want to devise proper marketing tactics to attract new customers, using location-based technology is an effective way to boost observational learning. By contrast, if the purpose is to retain existing customers, the role of observation learning in location-based networks should be minimal.

After discussing the causal social influence in location-based social media, we turn to non-causal mechanisms. In particular, Lee et al. (2016) examined homophily and network formation in location-based social networks. What are the main determinants of friendship creation in location-based social networks? Lee et al. (2016) used unstructured biography texts to build dyadic user similarity measures. Many social media platforms allow users to describe their interests in plain sentences. The issue is how we incorporate the unstructured text information and produce similarity metrics between users. Lee et al. (2016) applied Latent Dirichlet Allocation topic modeling to the text corpus of user biography texts. With a topic model, each user's biography can be presented as a topic vector. Each topic is an automatically generated user feature dimension that humans can easily understand. Then, they computed pairwise user similarity with the cosine similarity between topic vectors. Using 35 million check-in activities of 385,306 users at 3 million different locations worldwide, they found empirical evidence on the homophily effect in friendship creation of location-based social networks. Users who are more similar to each other in their unstructured biography texts are more likely to form friendships in location-based social media.

REFERENCES

Aral, S., Muchnik, L., & Sundararajan, A. (2009). Distinguishing influence-based contagion from homophily-driven diffusion in dynamic networks. *Proceedings of the National Academy of Sciences*, *106*(51), 21544–21549.

Banerjee, A. V. (1992). A simple model of herd behavior. *Quarterly Journal of Economics*, *107*(3), 797–817.

Bapna, R., Qiu, L., & Rice, S. (2017). Repeated interactions versus social ties: Quantifying the economic value of trust, forgiveness, and reputation using a field experiment. *MIS Quarterly*, *41*(3), 841–866.

Bapna, R., Ramaprasad, J., & Umyarov, A. (2018). Monetizing freemium communities: Does paying for premium increase social engagement? *MIS Quarterly*, *42*(3), 719–735.

Bapna, S., Benner, M. J., & Qiu, L. (2019). Nurturing online communities: An empirical investigation. *MIS Quarterly*, *43*(2), 425–452.

Berg, J., Dickhaut, J., & McCabe, K. (1995). Trust, reciprocity, and social history. *Games and Economic Behavior, 10*(1), 122–142.

Berger, J., & Milkman, K. L. (2012). What makes online content viral? *Journal of Marketing Research, 49*(2), 192–205.

Cabral, L., Ozbay, E. Y., & Schotter, A. (2014). Intrinsic and instrumental reciprocity: An experimental study. *Games and Economic Behavior, 87*, 100–121.

Cheng, H. K., Fan, W., Guo, P., Huang, H., & Qiu, L. (2020). Can "gold medal" online sellers earn gold? The impact of reputation badges on sales. *Journal of Management Information Systems, 37*(4), 1099–1127.

Chevalier, J. A., & Mayzlin, D. (2006). The effect of word of mouth on sales: Online book reviews. *Journal of Marketing Research, 43*(3), 345–354.

Chu, D. (2017). #CautionBusinesses: Using competitors' hashtags could possibly lead to trademark infringement. *Catholic University Journal of Law and Technology, 25*(2), 387–413.

Coy, P. (2012). The economics of Pussy Riot on YouTube. *Bloomberg Businessweek*, Last accessed on February9 , 2021, https://www.bloomberg.com/news/articles/2 012-09-21/the-economics-of-pussy-riot-on-youtube.

Ding, Y., Tu, Y., Pu, J., & Qiu, L. (2021). Environmental factors in operations management: The impact of air quality on product demand. *Production and Operations Management*, Forthcoming.

Gao, H., Kumar, S., Tan, Y. R., & Zhao, H. (2020). Socialize more, pay Less: Randomized field experiments on social pricing. Working paper, Available at SSRN: https://ssrn.com/abstract=3520844.

Ghose, A. (2018). *TAP: Unlocking the mobile economy.* MIT Press.

Gu, B., & Ye, Q. (2014). First step in social media: Measuring the influence of online management responses on customer satisfaction. *Production and Operations Management, 23*(4), 570–582.

Kumar, N., Venugopal, D., Qiu, L., & Kumar, S. (2018a). Detecting review manipulation on online platforms with hierarchical supervised learning. *Journal of Management Information Systems, 35*(1), 350–380.

Kumar, N., Qiu, L., & Kumar, S. (2018b). Exit, voice, and response on digital platforms: An empirical investigation of online management response strategies. *Information Systems Research, 29*(4), 849–870.

Kumar, N., Qiu, L., & Kumar, S. (2017). A hashtag is worth a thousand words: An empirical investigation of social media strategies in trademarking hashtags. *Mays Business School Research Paper* (2997653). Available at SSRN: https://ssrn.com/abstract=2997653.

Kwark, Y., Lee, G. M., Pavlou, P. A., & Qiu, L. (2021). On the spillover effects of online product reviews on purchases: Evidence from clickstream data. *Information Systems Research*, Forthcoming.

Lee, G. M., Qiu, L., & Whinston, A. B. (2016). A friend like me: Modeling network formation in a location-based social network. *Journal of Management Information Systems, 33*(4), 1008–1033.

Li, S., Luo, Q., Qiu, L., & Bandyopadhyay, S. (2020). Optimal pricing model of digital music: Subscription, ownership or mixed?. *Production and Operations Management, 29*(3), 688–704.

MacMillan, D., & Torbati, Y. (2021). How the rich got richer: Reddit trading frenzy benefited Wall Street elite. *The Washington Post*, Last accessed on February 9, 2021, https://www.washingtonpost.com/business/2021/02/08/gamestop-wallstreet-wealth/.

Narang, P. (2017). How brands are doing hashtag marketing? Last accessed on February 9, 2021, https://www.socialpilot.co/blog/how-brands-are-doing-hashtag-marketing.

Pu, J., Chen, Y., Qiu, L., & Cheng, H. K. (2020). Does identity disclosure help or hurt user content generation? Social presence, inhibition, and displacement effects. *Information Systems Research, 31*(2), 297–322.

Qiu, L., Tang, Q., & Whinston, A. B. (2015). Two formulas for success in social media: Learning and network effects. *Journal of Management Information Systems, 32*(4), 78–108.

Qiu, L., & Whinston, A. B. (2017). Pricing strategies under behavioral observational learning in social networks. *Production and Operations Management, 26*(7), 1249–1267.

Qiu, L., Shi, Z., & Whinston, A. B. (2018). Learning from your friends' check-ins: An empirical study of location-based social networks. *Information Systems Research, 29*(4), 1044–1061.

Victor, D., & Stevens, M. (2017). United Airlines passenger is dragged from an overbooked flight. *The New York Times*. Last accessed on February 9, 2021, https://www.nytimes.com/2017/04/10/business/united-flight-passenger-dragged.html.

Wang, L., Gopal, R., Shankar, R., & Pancras, J. (2015). On the brink: Predicting business failure with mobile location-based checkins. *Decision Support Systems, 76*, 3–13.

Wang, Z., Kumar, S., & Liu, D. (2020). On platform's incentive to filter fake reviews: A game-theoretic model. *International Conference on Information Systems*.

Wang, H., Du, R., Shen, W., Qiu, L., & Fan, W. (2021). Product reviews: A benefit, a burden, or a trifle? How seller reputation affects the role of product reviews, *MIS Quarterly*, Forthcoming.

Social Media, Fake Reviews, and Machine Learning Method

4

4.1 INTRODUCTION: FAKE NEWS AND FAKE REVIEWS

Misinformation is not new. The danger of misinformation was known to the ancient Chinese. As an ancient Chinese proverb says, "numerous mouths can melt metal, and accumulated slander can dissolve bones." A famous idiom about the virtuous statesman Zeng Shen and his trusting mother, "Zeng Shen committing murder," illustrates the power of misinformation vividly.

> There was a man who committed murder and had the same name as Zeng Shen. Somebody (mistakenly) informed Zeng Shen's mother, saying: "Zeng Shen committed murder." The mother continued weaving calmly. A short while later, another person informed her, saying: "Zeng Shen committed murder." The mother still weaved calmly. A short while later, yet another person informed her, saying: "Zeng Shen committed murder." The mother cast aside her weaving shuttle, pushed aside her loom, leapt across the walls, and fled. (Ng, 2020)

DOI: 10.1201/9781003196198-4

Even with Zeng Shen's virtue and his mother's trust in him, once the fake news was repeated three times, the mother became terrified. This idiom remains relevant in the age of social media in describing the impact of misinformation. Today what is different from ancient China is that social media platforms spread and repeat fake news faster and farther than ever before. Recent studies demonstrate an "illusory truth effect" for repetition: Even a statement is known to be false, it becomes less wild if people hear it again and again (Unkelbach et al., 2019). Social media platforms amplify the repetition of fake news dramatically. For example, on July 27, 2020, then-U.S. President Donald Trump tweeted a false claim made by a doctor: COVID-19 can be cured by hydroxychloroquine. A day later, tweets with similar content exploded at an average of over a million daily tweets (Sanders, 2021).

Therefore, social media platforms can be prone to abuse and manipulations, manifested in the form of fake news and fake reviews (Aral, 2020). Facebook and Twitter are found to be major social media platforms used to spread misinformation (Seetharaman et al., 2016). Furthermore, fake news outperforms real news in user engagement on social media platforms, and many people rate fake news headlines as somewhat or very accurate (Silverman & Singer-Vine, 2016). The consequence of such manipulations is not only the deterioration of information quality but also the spread of misinformation. For example, on September 25, 2020, a tweet accompanied by photos claims that more than 1,000 mail-in ballots had been discovered in a dumpster in Sonoma County in California. The truth is that the photos showed empty envelopes from the 2018 election that had been gathered for recycling. However, more than 25,000 Twitter users had shared this fake news story within a single day (Miller, 2020).

Besides fake news, business owners may manipulate their public online review ratings by using fake accounts or paid reviewers at Amazon or Yelp (Kumar et al., 2018). Several known cases of fake reviews have resulted in lawsuits (Roberts, 2015). The average cost of posting one five-star fake online product review can be as low as $1.29 (Stevens & Emont, 2018). Amazon initially allows sellers to generate reviews by sending free samples to incentivize customers to write online reviews, possibly because doing so can increase sellers' revenue and help Amazon earn profit. The incentivized reviews enable sellers to misrepresent the product quality (ReviewMeta, 2016). In October 2016, Amazon banned the incentivized reviews in order to build a reliable review system (Perez, 2016). To fight against fake reviews, social media platforms implement different policies to discourage review manipulations. TripAdvisor allows consumers to flag suspicious reviews. Amazon, Walmart, Yelp, and other e-commerce platforms try to tackle the problem by developing advanced algorithms to detect fake online product reviews (Stevens & Emont, 2018).

This chapter aims to pull back the curtain on social media misinformation and identify who spreads misinformation. Section 4.2 focuses on detecting and predicting fake reviewers operating on social media platforms using supervised and unsupervised machine learning methods. In Section 4.3, we examine sentiment manipulation as an operations activity in social media.

4.2 FAKE REVIEW DETECTION IN SOCIAL MEDIA

Consumers increasingly rely on online product reviews before making their purchase decisions (Kumar et al., 2018). Almost 70% of consumers trust the opinions expressed in online reviews (Rudolph, 2015). Therefore, sellers have strong incentives to manipulate their public online review ratings. Researchers show that a half-star increase in restaurant online review ratings at Yelp leads to a 19-percentage reduction in a restaurant's open reservations (Anderson & Magruder, 2012).

Distinguishing fake online reviewers from genuine users on social media platforms is challenging because fake online reviewers tend to mimic the behavior of genuine users. Based on a supervised learning approach, Kumar et al. (2018) proposed a novel approach to detect fake online reviewers. The intuition of their approach relies on a striking pattern shown in online review data. The features describing fake online reviewers are remarkably skewed in nature, which implies that fake online reviewers tend to distort the underlying natural distribution of online product reviews. Using a comprehensive dataset on online restaurant reviews from Yelp, Kumar et al. (2018) demonstrated the importance of modeling and identifying underlying distributional aspects of online reviewer behavior.

Kumar et al. (2018) first derived online reviewer features in their approach, including review gap, review count, rating entropy, rating deviation, time of review, and user tenure.

Review gap between successive reviews reflects the review-writing frequency of online reviewers. When all online reviews are posted within a short time frame, it indicates suspicious behavior.

Review count is the number of online reviews written by a particular reviewer. Paid online reviewers may have incentives to write more reviews than genuine online reviewers.

Rating entropy measures whether an online reviewer's rating scores are extreme. Fake online reviewers tend to post reviews with extreme rating scores because their goal is either to artificially inflate a restaurant's online review ratings or bring a bad reputation to its competitors. *Rating deviation* captures a user's deviation from the average product ratings. A genuine online reviewer's rating score is unlikely to differ from most online reviewers in every case dramatically.

Time of review is the average time difference between when a review is posted and when the very first review of that restaurant is posted. Fake online reviewers tend to post reviews extremely early after a restaurant's opening to maximize the impact of their reviews. For this reason, it should be a red flag if an online reviewer always posts restaurant reviews before any other reviewers.

User tenure is the amount of time an online reviewer is active on the social media platform. As an online reviewer is more active on the social media platform, the likelihood of being genuine increases. Fake online reviewers tend to have short-lived accounts and a high volume of reviews.

Kumar et al. (2018) first fit parametric distributions from univariate distribution families that best explain the empirical distribution for above features. Next, they considered the joint distribution of an online reviewer's rating and model it as a Dirichlet distribution. The logic is that even though individual reviews posted by fake online reviewers may look genuine at first glance, we can collectively capture anomalies in the online review patterns by modeling reviewer characteristics and interactions as univariate and multivariate distributions.

Kumar et al. (2018) adopted the filtered review mechanism at Yelp to categorize an online review as fake or genuine for training and test data set. They evaluated their approach using several supervised learning models, including logistic regression, support vector machine (SVM), AdaBoosting, and k-nearest neighbors (k-NN) with the proposed feature engineering approach. They show that using features with the proposed approach increases the likelihood of detecting fake online reviewers.

Besides the supervised learning approach, Kumar et al. (2019) adopted an unsupervised approach to detect anomalous behavior in online reviewers. A significant shortcoming of the supervised learning approach is the requirement of the labeled data (with fake reviewer/genuine reviewer labels). In general, it is hard to obtain labeled data, especially for fake online reviews. Similar to the supervised approach of Kumar et al. (2018), Kumar et al. (2019) learned probability distributions corresponding to univariate and multivariate features related to online reviewing behavior. However, after obtaining the empirical probability distributions, they developed an unsupervised approach to combine the distributions into a unified anomaly detection model.

These supervised and unsupervised approaches have direct implications for detecting fake online reviewers on social media platforms. Kumar et al. (2018, 2019) demonstrated that different features have varying levels of influence on the ability to detect fake online reviewers. With access to an ordered list of features, social media platforms and businesses can prioritize and learn about the most influential features of fake online reviewers and then incorporate appropriate interventions as a defensive mechanism against those fake online reviewers. Since the approaches of Kumar et al. (2018) and Kumar et al. (2019) provide an estimate about the probability of being an online reviewer based on the key features, social media platforms can develop a spam score for each online reviewer and share it with business owners and consumers.

Although the fake review problem is a challenge faced by all E-commerce platforms, they may not have an economic incentive to filter fake online product reviews (Pu et al., 2021; Wang et al., 2020). A striking finding of Wang et al. (2020) is that a more accurate algorithm of filtering fake online product reviews can sometimes lower E-commerce platforms' payoff. Therefore, the platforms do not always have an incentive to increase the accuracy of the algorithm, even if it is costless to do so.

4.3 SENTIMENT MANIPULATION AS AN OPERATIONS ACTIVITY IN SOCIAL MEDIA

Social media platforms have established themselves as powerhouses where consumers seeking information can interact among themselves and sellers, and sellers can actively identify target consumers and channel their advertisements accordingly. On these social media platforms, consumers can voice their opinions, and sellers can promote their products or services in various ways. In particular, sellers may strategically add positive sentiments themselves so that their products look better. Furthermore, because the goal of many social media platforms is to increase user participation and engagement, eradicating such sentiment manipulation is not necessarily a priority. Therefore, social media platforms are especially susceptible to sentiment manipulations, by which sellers deliberately create fake positive online reviews, comments, or likes. For instance, in a black market, Facebook "likes" can be purchased (Leonard, 2014). Twitter also suffers from the prevalence of fake tweets and automated programs (Coy, 2013).

Lee et al. (2018) looked at sentiment manipulation in the context of movie tweets. Some movie studios may use fake social media accounts or hire real social media accounts to post overly positive messages about their movies and attract consumers. For example, Sony was fined by the Connecticut attorney general for creating fake reviews for at least four of its movies (Zielbauer, 2002). The phenomenon of sentiment manipulation in the movie industry is neither new nor exclusive to the U.S. Sina Weibo (often described as "China's Twitter") has been used by opportunistic companies, and the movie industry has quickly adopted sentiment manipulation for promotion and advertising in China. Paid posting on Sina Weibo is a well-managed activity by public relations companies involving thousands of individuals and tens of thousands of different online accounts. These fake accounts or hired accounts are called hidden paid posters or termed "Internet water army" in China. A quality control team checks whether fake comments meet a certain "quality" threshold. For instance, a comment would not be validated if it is deleted by the host (Zhou, 2012). The Chinese movie "The Last Supper" admitted hiring an "Internet water army" to raise its online rating on review websites and endorse the movie on Sina Weibo. Some insiders charge that using a water army as part of a movie's online promotion is already widely known in the Chinese movie industry: Many movies resort to sentiment manipulation, each at the cost of over 1 million Chinese yuan (Zhou, 2012).

To understand movie studios' incentives to conduct sentiment manipulation, we need to dig deep into the major players in the movie industry: producers, distributors, and exhibitors (McKenzie, 2012). A producer may buy a screenplay, buy a book to adapt into a screenplay, or hire a writer to develop an idea, and then make a movie. A distributor distributes the movie; it also makes important operational decisions, such as choosing a release date and designing and implementing an advertising campaign. Finally, exhibitors are movie theaters that show movies to audiences. There has been a trend of vertical integration in the movie industry: Movie studios increasingly both produce and distribute movies themselves (Gilchrist & Sands, 2016). Movie producers and distributors generally act like integrated firms. Therefore, we regard a movie studio as both a producer and distributor.

One of the most important decisions of a movie studio is an effective advertising campaign (McKenzie, 2012), and sentiment manipulation can be a promotion strategy for movie studios. Movie studios might be incentivized to conduct more sentiment manipulation prior to the movie release and less afterward because of the following two reasons.

First, movie studios typically charge fees as percentages of box office revenues rather than a flat fee. The share division of a movie studio changes over weeks of the movie's run, with a smaller share for the movie studio in later weeks. For a movie, it is standard for the movie studio to keep as much as 90% of revenues in the opening week. Hence, the incentive for the movie studio to conduct sentiment manipulation should be high before the opening weekend release. After the opening week, the distributor's share drops dramatically to 50% or even 30% (Gilchrist & Sands, 2016).

Second, in the movie industry, it is widely believed that the opening weekend is critical for movie studios. A movie that fails to open strongly almost always loses the attention of the media, audiences, and exhibitors. According to Box Office Mojo, the opening weekend accounts for a significant fraction of a film's box office, typically 30%–45% (Lee et al., 2018). Additionally, if the opening weekend box office revenue is low, the movie theaters may drop the movie or reduce the number of screens. In general, "a film's opening weekend is usually the most lucrative one for its studio. Financial agreements with theaters normally give the filmmaker a greater percentage of the box office during the first weeks of release. And in this glutted market, studio executives also worry that theaters will replace a film with another if it doesn't win audiences quickly" (Corts, 2001, p. 514). Therefore, the incentive of movie studios to conduct sentiment manipulation drops significantly after the release.

Compared with other professional movie review platforms, Twitter is a relatively open platform, making it much easier for movie studios to engage in sentiment manipulation. Specifically, movie studios can easily manufacture tweets that appear to be posted by individual consumers to influence consumers' movie-going decisions. While Twitter is perhaps not a major platform that consumers would visit in search of product reviews, many companies today adopt tweeting as a new marketing tool, as discussed in Chapter 3. In reality, people may follow Twitter trends or movie studio's official Twitter accounts to obtain the latest movie news, and movie studios can use fake Twitter accounts or hire real Twitter accounts to post overly positive messages about their movies and attract customers.

Lee et al. (2018) compiled a list of 482 movies released in the United States in the years 2012 and 2013. For each movie, they used its corresponding hashtag (#) to identify and collect its tweets that are written in English, starting 60 days prior to its release day and up to 60 days after the release day. They then used the trained Naive Bayes classifier to classify the polarity of these tweets and aggregated them daily to construct a daily average Twitter sentiment measure for each movie, with zero being the lowest possible sentiment level and one the highest. If there is sentiment manipulation on Twitter, we should observe a significant drop in the sentiment level after the release day because, as mentioned earlier, the incentive of movie studios to conduct sentiment manipulation should be high before the movie release, and then drop significantly after the release.

The analysis of Lee et al. (2018) shows that, after controlling for other factors that may affect movie sentiment, the average sentiment and the proportion of highly positive tweets both exhibit a significant drop on the movie's release day, which suggests the potential existence of sentiment manipulation on Twitter. They also find that major studio movies tend to have a smaller drop in Twitter sentiment than non-major studio movies. It means that a major studio production is less likely to engage in manipulative behavior because of reputation concerns, whereas independently produced movies have fewer reputation concerns.

Furthermore, Lee et al. (2018) examined how competition impacts movie studio's sentiment manipulation. They considered competition measures based on the following two dimensions: timing competition and thematic competition. Timing competition is defined as the time interval between these two movies' release days. If two movies are released simultaneously, they are likely to be competing with each other directly. Thematic competition is characterized by how similar any given pair of movies are to each other. Lee et al. (2018) employed a machine learning technique, topic modeling, and used movie keywords as inputs to the topic model to uncover the underlying topic distributions of each movie. Specifically, they used latent Dirichlet allocation (LDA), a type of topic modeling method used extensively in text mining tasks, to discover latent topics. They then computed the cosine similarity between a pair of movies' relative topic distributions as our measure of thematic similarity between these two movies. Based on these competition measures, they find that movies facing a higher level of competition have a larger drop in Twitter sentiment than those facing a lower level of competition, which suggests that movie studios manipulate more when competition increases. It echoes research in some other contexts: Fake reviews are more prevalent for hotels that face intense competition (Mayzlin et al., 2014). Although competition is often praised as a driving force for innovation (Chen et al., 2021), it is not always beneficial.

We have seen so far that the observed Twitter sentiment patterns are consistent with sentiment manipulation, and the manipulation level is higher when there is fiercer competition. Next, we investigate movie studio's positive sentiment manipulation (i.e., inflate their own movies' sentiment) and negative sentiment manipulation (i.e., bring a bad reputation to its competitors' movies). In the prior academic literature and business practice, little is known about the relative prevalence of positive and negative manipulation. On the one hand, the benefit of conducting negative manipulation is typically higher than conducting positive manipulation. A negative review hurts more than a positive review helps (Chevalier & Mayzlin, 2006). If a negative tweet has a greater role in affecting consumer purchase, other things being equal, a movie studio will have a greater incentive to conduct negative manipulation than positive manipulation given the manipulation budget.

On the other hand, the cost of conducting negative manipulation is higher than conducting positive manipulation. "Because giving negative comments usually brings side effects and has risks, few Internet public relations companies will take such orders. So in most cases, film companies will use staff they trained themselves for negative comments about opponents" (Zhou, 2012). Even if some public relations companies are willing to conduct negative manipulation, they would be more cautious. To avoid problems, they will often register new accounts from a foreign Internet protocol (IP) address. Training the internal staff or registering new accounts from a foreign IP address involves additional costs. As a result, both the benefit and cost of conducting negative manipulation are higher than those of conducting positive manipulation. Without empirical evidence, it is not clear whether positive manipulation is more prevalent than negative manipulation. The results of Lee et al. (2018) inform that positive sentiment manipulation is much more common than negative sentiment manipulation in the context of movies' tweets.

In summary, the above studies show that the observed Twitter sentiment exhibits patterns consistent with sentiment manipulation. In addition, the manipulation level is higher as the competition intensifies. These results can be generalized to businesses outside the movie industry. However, it should be noted that while the quality of a movie is mainly time-invariant, that of hotels, restaurants, and medical practices can be time-variant. The implication is that while movie studios may primarily conduct sentiment manipulation before the movie release, hotels, restaurants, and medical practices might find the need to engage in strategic sentiment manipulation regularly.

REFERENCES

Anderson, M., & Magruder, J. (2012). Learning from the crowd: Regression discontinuity estimates of the effects of an online review database. *The Economic Journal, 122*(563), 957–989.

Aral, S. (2020). *The hype machine: How social media disrupts our elections, our economy, and our health-and how we must adapt.* New York: Currency.

Chen, T., Cheng, H. K., Jin, Y., Li, S., & Qiu, L. (2021). Impact of competition on innovations of IT industry: An empirical investigation, *Journal of Management Information Systems*, Forthcoming.

Chevalier, J. A., & Mayzlin, D. (2006). The effect of word of mouth on sales: Online book reviews. *Journal of Marketing Research, 43*(3), 345–354.

Corts, K. S. (2001). The strategic effects of vertical market structure: Common agency and divisionalization in the US motion picture industry. *Journal of Economics & Management Strategy, 10*(4), 509–528.

Coy, P. (2013). Could bots and spam smother the Twitter IPO? *Bloomberg Businessweek*, Last accessed on February 17, 2021, https://www.bloomberg.com/news/articles/2013-09-25/could-bots-and-spam-smother-the-twitter-ipo.

Gilchrist, D. S., & Sands, E. G. (2016). Something to talk about: Social spillovers in movie consumption. *Journal of Political Economy*, *124*(5), 1339–1382.

Kumar, N., Venugopal, D., Qiu, L., & Kumar, S. (2018). Detecting review manipulation on online platforms with hierarchical supervised learning. *Journal of Management Information Systems*, *35*(1), 350–380.

Kumar, N., Venugopal, D., Qiu, L., & Kumar, S. (2019). Detecting anomalous online reviewers: An unsupervised approach using mixture models. *Journal of Management Information Systems*, *36*(4), 1313–1346.

Lee, S. Y., Qiu, L., & Whinston, A. (2018). Sentiment manipulation in online platforms: An analysis of movie tweets. *Production and Operations Management*, *27*(3), 393–416.

Leonard, A. (2014). Facebook's black market problem revealed. *Salon*, Last accessed on February 17, 2021, https://www.salon.com/2014/02/14/facebooks_big_like_problem_major_money_and_major_scams/.

Mayzlin, D., Dover, Y., & Chevalier, J. (2014). Promotional reviews: An empirical investigation of online review manipulation. *American Economic Review*, *104*(8), 2421–2455.

McKenzie, J. (2012). The economics of movies: A literature survey. *Journal of Economic Surveys*, *26*(1), 42–70.

Miller, G. (2020). As U.S. election nears, researchers are following the trail of fake news. *Science News*, Last accessed on February 17, 2021, https://www.sciencemag.org/news/2020/10/us-election-nears-researchers-are-following-trail-fake-news.

Ng, A. (2020). Making sense of the pandemic with classical Chinese idioms. *British Journal of Chinese Studies*, *10*. 10.51661/bjocs.v10i0.113.

Perez, S. (2016). Amazon bans incentivized reviews tied to free or discounted products. Last accessed on February 17, 2021, https://techcrunch.com/2016/10/03/amazon-bans-incentivized-reviews-tied-to-free-or-discounted-products/.

Pu, J., Nian, T., Qiu, L., & Cheng, H. K. (2021). Platform policies and sellers' competition in agency selling in the presence of online quality misrepresentation. *Journal of Management Information Systems*, Forthcoming.

ReviewMeta (2016). Analysis of 7 million Amazon reviews: Customers who receive free or discounted item much more likely to write positive review. Last accessed on February 17, 2021, https://reviewmeta.com/blog/analysis-of-7-million-amazon-reviews-customers-who-receive-free-or-discounted-item-much-more-likely-to-write-positive-review/.

Roberts, J. (2015). Amazon sues people who charge 5 for fake reviews. *Fortune Magazine*, Last accessed on February 17, 2021, http://fortune.com/2015/10/19/amazon-fake-reviews.

Rudolph, S. (2015). The impact of online reviews on customers' buying decisions, *Business 2 Community*, Last accessed on February 17, 2021, http://www.business2community.com/infographics/impact-online-reviewscustomers-buying-decisions-infographic-01280945#iZwM69pSgVKLlH6A.97.

Sanders, M. (2021). A few simple tricks make fake news stories stick in the brain. Last accessed on February 17, 2021, https://www.sciencenews.org/article/misinformation-fake-news-stories-social-media-brain.

Seetharaman, D., Nicas, J., & Olivarez-Giles, N. (2016). Social-media companies forced to confront misinformation and harassment. *The Wall Street Journal*, Last accessed on February 17, 2021, https://www.wsj.com/articles/social-media-companies-forced-to-confront-misinformation-and-harassment-1479218402.

Silverman, C., & Singer-Vine, J. (2016). Most Americans who see fake news believe it, new survey says. *BuzzFeed News*, Last accessed on February 17, 2021, https://www.buzzfeednews.com/article/craigsilverman/fake-news-survey.

Stevens, L., & Emont, J. (2018). How sellers trick Amazon to boost dales. *The Wall Street Journal* (July 28), Last accessed on February 17, 2021, https://www.wsj.com/articles/how-sellers-trick-amazon-to-boost-sales-1532750493.

Unkelbach, C., Koch, A., Silva, R. R., & Garcia-Marques, T. (2019). Truth by repetition: Explanations and implications. *Current Directions in Psychological Science*, *28*(3), 247–253.

Wang, Z., Kumar, S., & Liu, D. (2020). On platform's incentive to filter fake reviews: A game-theoretic model. *International Conference on Information Systems*.

Zielbauer, P. (2002). Metro briefing I Connecticut: Hartford: Falsified Movie Reviews. *The New York Times*, Last accessed on February 17, 2021, https://www.nytimes.com/2002/03/13/nyregion/metro-briefing-connecticut-hartford-falsified-movie-reviews.html.

Zhou, R. (2012). Muddy waters. Last accessed on February 17, 2021, http://usa.chinadaily.com.cn/life/2012-12/13/content_16013662.htm.

Social Media and Healthcare 5

5.1 INTRODUCTION: ONLINE HEALTH COMMUNITIES

On March 11, 2020, COVID-19 was declared a pandemic by the World Health Organization (WHO) due to its fast spread worldwide. Online health communities have numerous benefits to offer during this pandemic, such as improved protection for vulnerable communities and increased access to benefits for populations that predominantly stay indoors. The COVID-19 pandemic caused a significant disruption in the offline healthcare channel, and online health communities illustrate the digital resilience in recovering from and adjusting to this massive exogenous disruption.

In recent years, with the rapid development of information technologies, online health communities with social media features are becoming increasingly important in supporting access to the mass of information and resources and promoting barrier-free communication and information exchanges between physicians and patients (Wang et al., 2020; Yan & Tan, 2014). Although physicians have long recognized the importance of incorporating patient-specific information into treatment plans, they often have little information about patients' daily life styles (Chen et al., 2021). Online health communities provide an opportunity for interactions between physicians and patients.

In reality, patients and physicians are rapidly embracing online health communities. Recent studies have documented the efficacy of online health communities as a viable tool to enrich patients' knowledge of healthcare, facilitate healthcare decision-making, and improve the doctor–patient relationship (Huang et al., 2019). However, there is no prevailing wisdom on the best design of online healthcare communities. In Section 5.2, we look at interactions between physicians and patients in healthcare question-and-answer (Q&A) forums. We examine how online interactions between physicians and patients affect offline appointments in Section 5.3.

5.2 ONLINE PHYSICIAN RESPONSES AND SOCIAL MEDIA

As a form of social media website, Q&A forums have steadily gained popularity because of the quality content hosted on these sites. As user interest rises, Q&A sites are becoming increasingly domain-specific and niche. In knowledge- or expertise-intensive specialist services, for example, physician selection, quality content hosted on Q&A forums is of interest to physicians and patients.

The primary healthcare market is regulated in most countries. The delivery of healthcare services is governed by local laws that discourage or prohibit promotions by physicians. Quality responses from physicians on Q&A forums serve as valuable signals of physicians' expertise. The healthcare market is full of asymmetric information. Physicians have more information than patients. As a result, physicians can recommend unnecessary care that enhances their incomes even though it may not benefit patients. Because of their limited knowledge, patients have no reliable way of evaluating the quality of the advice they are getting. Even after the medical service is delivered, patients still may not be able to evaluate the quality of the service.

Therefore, physicians' online responses become an important way to reduce information asymmetry. Doctors use thoughtful online responses not only to socially interact with patients but also to signal their expertise. If a high-expertise doctor can credibly convey quality information by providing thoughtful online responses, then the doctor is likely to receive more recommendations from patients. The informational value of physicians' online responses relies on whether high-expertise doctors can separate themselves from low-expertise doctors.

Khurana et al. (2019) investigated the impact of physicians' responses to patients' questions on recommendations for responding doctors. They seek to understand whether introducing physicians' response feature in online healthcare portals impacts the satisfaction level of patients. They show the benefit of introducing a Q&A feature from the viewpoint of all stakeholders. For patients, the Q&A feature provides an additional distinguishing variable in selecting a physician. After launching the feature of physicians' online responses, Khurana et al. (2019) show a small increase in the level of patient recommendations for physicians who can respond to patients' questions but choose not to do so. However, physicians who actually respond to patients' questions enjoy a much higher level of recommendations than before. With the patient recommendations driving business, physicians have incentives to respond to patients' questions.

Given the importance of physicians' responses and engagement we have already seen, online healthcare portals develop new features to encourage health-related knowledge sharing and incentivize physicians' online responses. One is allowing patients to send digital gifts to doctors who have helped physicians as a token of appreciation. Some online healthcare portals have integrated online payment services and introduced digital monetary gifts to community members beyond traditional free and nonmonetary gifts (Wang et al., 2020).

Wang et al. (2020) examined the impact of introducing monetary gifts on the responses of a focal physician in an online healthcare portal. On the one hand, we may expect that monetary gifts, compared with nonmonetary gifts, can better motivate physicians to participate in an online healthcare portal and contribute their professional expertise. The monetary gift feature may promote the physician-patient relationship and encourage physicians to help patients better. By giving more valuable monetary gifts, patients try to show appreciation and express gratitude at higher levels. Physicians who receive such gifts can undoubtedly capture the emotion of thankfulness and feel more recognized for knowledge sharing. Therefore, physicians can be more attentive to help patients and respond more actively to their medical consultations.

However, on the other hand, the introduction of monetary gifts may also affect physicians' enthusiasm negatively for delivering online responses to medical consultations. In particular, many physicians are voluntarily using their spare time to help patients online. Extra monetary rewards from patients may make physicians more inclined to regard online consultation as a paid job, thus weakening their selfless motivation to help others using their professional knowledge. Monetary gifts can introduce concerns about social reputation or self-respect and crowd out the motivations to volunteer (Burtch et al., 2018). This is known as the crowding-out effect, where external treatment modifies the motivation of contribution. Consequently, both directions are theoretically plausible and practically ambiguous, motivating Wang et al. (2020) to empirically examine the impact of providing informal payments on physicians' responses to medical consultations.

Wang et al. (2020) leveraged the introduction of the monetary gift feature by a leading online healthcare portal in China as a natural experiment that exogenously provides physicians with extra monetary rewards. They find that introducing informal payments into the online healthcare portal reduces the number of physicians' online responses, indicating a crowding-out effect of extra monetary reward from patients on physicians' intrinsic motivation of contribution.

This striking result implies that providing extra monetary rewards tends to affect physicians' enthusiasm for delivering online responses to medical consultations. As many physicians voluntarily use their spare time to help patients online, informal payments can make them more inclined to regard online consultation as a paid job, crowding out their intrinsic motivations to volunteer. Furthermore, accepting informal payments from patients may bring concerns about physicians' reputation as the public usually expects the physician community to be caring, competent, and altruistic. Before launching a new feature and introducing extra rewards, managers of online healthcare portals should consider whether introducing monetary incentives may crowd out the intrinsic motivations. Although digital gifting practices have been implemented in other types of online communities to establish, sustain, and promote social connections, the results of Wang et al. (2020) suggest that online health communities should be wary of the potential adverse effects.

5.3 ONLINE CONSULTATION AND OFFLINE APPOINTMENTS

Online healthcare portals play an essential role in broadening and diversifying the available channels to serve patients. As discussed earlier, one common form that a healthcare portal takes is the online Q&A website, which provides a bridge between patients and doctors and reduces the time and cost for patients when they are seeking healthcare services.

Fan et al. (2020) studied whether patients' behavior of choosing doctors to consult offline is affected by the performance of physicians' online consultation services. On the one hand, a good physician can credibly convey their expertise and responsibility by providing online consultations. Therefore, physicians' online consultation service may increase their offline appointments. On the other hand, one may also argue that the cannibalization effect may exist. If a physician successfully answers patients' questions online, the patients may not need to make offline appointments.

The dataset in Fan et al. (2020) was collected from Haodf.com, a major third-party online Q&A website in China. The healthcare platform allows patients to ask questions about their illnesses to physicians, who can then answer them via an online channel. Fan et al. (2020) find that physicians who open the online consultation service generate more offline appointments than those who do not. Essentially, opening online consultation can improve the interaction between physicians and patients and provide better opportunities for physicians to communicate with patients who are far away. Hence, physicians should pay attention to the new features launched by online healthcare portals. Furthermore, for online healthcare portals, the benefits of providing online Q&A cannot be ignored: The feature of online consultations can attract more physicians and patients.

REFERENCES

Burtch, G., Hong, Y., Bapna, R., & Griskevicius, V. (2018). Stimulating online reviews by combining financial incentives and social norms. *Management Science*, *64*(5), 2065–2082.

Chen, W., Lu, Y., Qiu, L., & Kumar, S. (2021). Designing personalized treatment plans for breast cancer. *Information Systems Research*, Forthcoming.

Fan, W., Zhou, Q., Qiu, L., & Kumar, S. (2020). Should doctors open online consultation? An empirical investigation of how it impacts the number of offline appointments. Working paper, Available at SSRN: https://ssrn.com/abstract=3 622761.

Huang, K. Y., Chengalur-Smith, I., & Pinsonneault, A. (2019). Sharing is caring: Social support provision and companionship activities in healthcare virtual support communities. *MIS Quarterly*, *43*(2), 395–424.

Khurana, S., Qiu, L., & Kumar, S. (2019). When a doctor knows, it shows: An empirical analysis of doctors' responses in a Q&A forum of an online healthcare portal. *Information Systems Research*, *30*(3), 872–891.

Wang, Q., Qiu, L., & Xu, W. (2020). Informal payments and doctor engagement in online health community: An empirical investigation using generalized synthetic control. Working paper, Available at SSRN: https://ssrn.com/abstract=3691702.

Yan, L., & Tan, Y. (2014). Feeling blue? Go online: An empirical study of social support among patients. *Information Systems Research*, *25*(4), 690–709.

Social Media and Telecommunications

6

6.1 INTRODUCTION: SOCIAL MEDIA AND 5G

As smartphones become the primary screen for consumers, mobile data usage per consumer has been growing dramatically. In the first quarter of 2019, mobile devices generated 48.71% of global website traffic, consistently hovering around the 50% mark since the beginning of 2017 (Clement, 2020). According to a recent report, the total number of global mobile subscribers will grow from 5.1 billion in 2018 to 5.7 billion by 2023 (Cisco, 2020). We can foresee continuous massive growth in mobile data consumption in the era of 5G, where we have higher connection speed and more streaming music, on-demand videos, and high-definition gaming on our mobile devices. For example, 90% of the feeds on social media platforms, such as Facebook, YouTube, and WhatsApp, are videos (Sengupta, 2019).

With the upcoming 5G Internet, the Internet of Things (IoT) connects billions of physical devices worldwide, collecting and sharing data for smart cars, smart homes, and smart cities. This fast-growing market provides a huge opportunity for content providers (e.g., Facebook, Netflix, and Spotify) as well as mobile service providers (e.g., AT&T, Verizon, and T-Mobile).

Given this tremendous increase in data consumption by consumers and the upcoming next-generation 5G networks, the telecommunications industry has been exploring new business models. On the one hand, to manage the increasing mobile data traffic, mobile service providers experiment with third-party Wi-Fi hot spots to augment their cellular capacity (Qiu et al., 2019).

DOI: 10.1201/9781003196198-6

On the other hand, mobile service providers collaborate with other industry partners, such as content providers and advertisers, to better deliver their services to consumers (Cho et al., 2016; Cho et al., 2020; Mei et al., 2021).

This chapter aims to examine the new business models inspired by social media and telecommunications technology. Section 6.2 introduces a new business model proposed by mobile service providers, sponsored data program, which transfers a fraction of the data bill from consumers to content providers. In Section 6.3, we examine an emerging monetization mechanism: Sponsored data with reward tasks. Under this mechanism, consumers are subsidized with free megabytes by content providers in exchange for engagement with advertisers by performing various forms of reward tasks on social media sites.

6.2 SPONSORED DATA PROGRAM

The industry practitioners, both the mobile service providers and the content providers, benefit from high revenue due to the high demand for mobile data. To encourage premium content consumption, innovative content providers, such as Netflix and Spotify, and social media websites are producing more and more content in creative ways. While enjoying various premium content provided by content providers, consumers, however, are becoming more conscious about their data consumption since their monthly caps of mobile data plans can be easily exhausted by premium content, such as high-definition videos and virtual reality (VR) or augmented reality (AR) games.

In view of this, mobile service providers have proposed a business model to transfer a fraction of the data bill from consumers to content providers. In their 2014 Developer Summit, the executives of AT&T introduced "Sponsored Data," which allows customers to browse, stream, and enjoy content from their data sponsors without impacting their monthly data plan allowance. This new monetization mechanism was quickly embraced by the industry, with ten companies signed up with AT&T one year after their proposal. T-Mobile started offering its customers free streaming music from top providers, including Spotify, Pandora, iTunes Radio, in 2014. In November 2015, free streaming video was provided through a new service called "Binge On," which cooperates with 42 providers – Netflix, Amazon, Hulu, HBO, among others. In early 2016, Verizon introduced its sponsored data service FreeBee Data. A black-and-white bee appears next to sponsored content, so customers know that clicking on that content does not incur data charges (Qiu et al., 2016). Under these similar sponsored data arrangements, eligible data usage charges are billed directly to the sponsoring company (typically the content providers). Data sessions are identified by a "sponsored data" icon on the end user's device, and the usage is itemized separately on the end user's monthly invoice. Essentially, this new business model allows content providers to pay for the mobile data that end users consume.

While sponsored data is adopted as a new business model by most major U.S. carriers and applauded by consumers, such a business model has been criticized by advocators of net neutrality (Cheng et al., 2011). The Federal Communications Commission (FCC) enforced net neutrality by issuing the "open Internet order" and barred service providers from discriminating against certain types of traffic or creating pay-to-play fast lanes (Qiu et al., 2016). Supporters of net neutrality argue that such arrangements contradict the spirit of net neutrality, meaning that all content transmitted over the Internet should be treated equally. They argue that sponsored data plans treat the sponsored content preferentially because consumers have more incentive to consume "free" packets over unsubsidized ones. For example, Facebook's Free Basics platform provides free Internet to a limited number of websites in developing countries through local service providers. India's regulators banned Free Basics because according to them, it violates the principles of net neutrality (Qiu et al., 2016). The biggest objection is that the Free Basics platform offers only a few content providers chosen and controlled by Facebook.

However, mobile service providers argue that this new business model does not violate the principle of net neutrality since sponsored content is transmitted without priority over non-sponsored content. U.S. regulators are exploring the welfare implications of the new sponsored data plans. In letters to AT&T and T-Mobile, the FCC wrote, "We want to ensure that we have all the facts to understand how these services (sponsored data) relate to the commission's goal of maintaining a free and open Internet while incentivizing innovation and investment from all sources" (Kang, 2015).

Cho et al. (2016) analyzed the incentives of mobile service providers, content providers, and consumers under sponsored data plans. They find that sponsored data can force the content providers into a bidding war like a classic "prisoner's dilemma." Content providers would prefer not to sponsor consumers' mobile data, but they know that if they do not sponsor, others will do and drive their rivals out of the market. At first glance, sponsored data plans would benefit consumers because they allow users to consume mobile data for free. However, Cho et al. (2016) show that the variety of content may be reduced, which in the long run harms consumers. The reason is that under sponsored data, small content providers and developers are put at a distinct disadvantage to their deeper-pocketed competitors, who can make their content more easily accessible by paying to exempt their traffic from consumers' monthly bills. More importantly, mobile service providers could pick winners and losers using their position as a gatekeeper, which would distort competition of content providers and hurt innovations on the Internet in the long run.

The above policy concern could be more serious when a mobile service provider tries to acquire a content provider, which is termed vertical integration (Cho et al., 2020). AT&T's historical acquisition of Time Warner represents the mobile service provider's most ambitious bid to compete with technology companies, such as Netflix and Amazon. The U.S. Justice Department had deep concerns that the merger could hamper competition among content providers and ultimately harm consumers via higher prices and restricted access to certain content. Cho et al. (2020) demonstrated that the combination of vertical integration and sponsored data could put the prominent, integrated content provider at a unique advantage over small content providers.

6.3 REWARD TASKS AND SOCIAL MEDIA

Recently, a new monetization mechanism, reward task, also known as incentivized ads, came into public notice and is gaining popularity among industry practitioners (Guo et al., 2019). This innovative business model puts a more positive spin on ad formats and offers a benefit to customers, for example, free megabytes of data credited to the data cap, in exchange for engagement with the advertisers by performing some tasks on social media sites, such as viewing social media ads, downloading social media apps, and shopping through social media apps. Therefore, it is a combination of traditional advertising and sponsored data discussed earlier.

The telecommunications industry has adopted this novel business model. The solution provider Datami has already developed close partnerships with tier-one network carriers across six continents, serving more than 1.3 billion consumers, helping their clients scale their advertising, while increasing the average revenue per user (ARPU), and accelerating monetization of new digital services. According to Radeke (2017), among the top 25 U.S. non-gaming media sources that were ranked in The AppsFlyer Performance Index in 2017, more than half list some sort of reward task format in the portfolio.

The content provider's decision of the optimal subsidization rate (how much free mobile data in exchange for viewing ads) lies in the center of the business model of reward tasks. With the upcoming 5G era, where people's mobile data packages can be easily chewed up by a few high-definition videos or games, designing sponsored data plans that are attractive to consumers poses even more challenges for the content provider. Kumar et al. (2020a) find that the optimal design of reward tasks is strikingly different from traditional advertising, which is discussed in previous studies (Kumar, 2015; Kumar et al., 2020b; Liu et al., 2020). At first glance, when the revenue rate of reward tasks is higher, the content provider's optimal subsidization rate should also be higher. A startling discovery of Kumar et al. (2020a) is that the content provider's optimal subsidization rate decreases in the revenue rate of reward tasks. The reason is the strategic action of the mobile service provider: Although a higher revenue rate of reward tasks could potentially generate more profit for the content provider, more surplus is actually extracted by the service provider through charging the content provider a higher price for sponsored data. As a result, the content provider becomes better off by lowering the subsidization rate. This striking result helps content providers decide their optimal subsidization rates based on their revenue rates of advertising in the design of reward tasks.

Furthermore, the analysis of Kumar et al. (2020a) reveals that introducing sponsored data and reward tasks from the content provider might not necessarily increase consumer surplus. Specifically, when the content provider's revenue rate of traditional advertising is lower than that of reward tasks, consumer surplus is smaller than under the scenario where there are no sponsored data and reward tasks. Therefore, the new business practice of rewarding consumers with sponsored data after they accomplish a task is not always welfare-enhancing for customers. Thus, although major mobile service providers claim this new business practice to be welfare-enhancing, it requires attention from policymakers to investigate the detailed market conditions.

REFERENCES

Cheng, H. K., Bandyopadhyay, S., & Guo, H. (2011). The debate on net neutrality: A policy perspective. *Information Systems Research*, 22(1), 60–82.

Cho, S., Qiu, L., & Bandyopadhyay, S. (2016). Should online content providers be allowed to subsidize content?—An economic analysis. *Information Systems Research*, 27(3), 580–595.

Cho, S., Qiu, L., & Bandyopadhyay, S. (2020). Vertical integration and zero-rating interplay: An economic analysis of ad-supported and ad-free digital content. *Journal of Management Information Systems*, 37(4), 988–1014.

Cisco (2020). Cisco annual Internet report (2018–2023) white paper. Last accessed on February 17, 2021, https://www.cisco.com/c/en/us/solutions/collateral/executive-perspectives/annual-internet-report/white-paper-c11-741490.html.

Clement, J. (2020). Global mobile data traffic from 2017 to 2022. *Statista*. Last accessed on February 17, 2021, https://www.statista.com/statistics/271405/global-mobile-data-traffic-forecast/.

Guo, H., Zhao, X., Hao, L., & Liu, D. (2019). Economic analysis of reward advertising. *Production and Operations Management, 28*(10), 2413–2430.

Kang, C. (2015). F.C.C. asks Comcast, AT&T and T-Mobile about "zero-rating" services. *New York Times*, Last accessed on February 17, 2021, http://bits.blogs.nytimes.com/2015/12/17/f-c-c-asks-comcast-att-and-t-mobile-about-zero-rating-services/?_rD0.

Kumar, S. (2015). *Optimization issues in web and mobile advertising: Past and future trends*. New York: Springer.

Kumar, S., Mei, X., Qiu, L., & Wei, L. (2020a). Watching ads for free mobile data: A game-theoretic analysis of sponsored data with reward task. Available at SSRN: https://ssrn.com/abstract=3640368.

Kumar, S., Tan, Y., & Wei, L. (2020b). When to play your advertisement? Optimal insertion policy of behavioral advertisement. *Information Systems Research, 31*(2), 589–606.

Liu, D., Kumar, S., & Mookerjee, V. S. (2020). Flexible and committed advertising contracts in electronic retailing. *Information Systems Research, 31*(2), 323–339.

Mei, X., Cheng, H. K., Bandyopadhyay, S., Qiu, L. , & Wei, L. (2021). Sponsored data: Smarter data pricing with incomplete information. *Information Systems Research*, Forthcoming.

Qiu, L., Cho, S., & Bandyopadhyay, S. (2016). Net neutrality may be at risk when companies like Netflix subsidize your data. *Conversation*, Last accessed on February 17, 2021, https://theconversation.com/net-neutrality-may-be-at-risk-when-companies-like-netflix-subsidize-your-data-56049.

Qiu, L., Rui, H., & Whinston, A. (2019). Optimal auction design for Wi-Fi procurement. *Information Systems Research, 30*(1), 1–14.

Radeke, J. (2017). Reward advertising in 2017: The good, the bad & the ugly. *AppsFlyer*. Last accessed on February 17, 2021, https://www.appsflyer.com/rewarded-advertising-2017-good-bad-ugly/.

Sengupta, D. (2019). Consumption to double: Data usage to turn upwardly mobile in 2019. *The Economic Times*. Last accessed on February 17, 2021, https://economictimes.indiatimes.com/tech/internet/data-usage-to-turn-upwardly-mobile-in-2019/articleshow/67358232.cms?from=mdr.

Future Trends and Challenges in Social Media

7

7.1 SOCIAL TRADING AND FINTECH: INDIVIDUAL INVESTORS IN THE AGE OF SOCIAL MEDIA

In this chapter, we discuss some of the key future trends in social media and associated challenges. One such future trend is social trading. In the past, individual investors mostly carried out trades based on publicly available information about a stock as well as their own private information, while a few expert traders would try to surmise the opinion of the crowd. With the advent of Fintech in trading markets, many technology startups are using social media analytics to gauge investor sentiment. Some of them are socially-enabled trading platforms that specialize in aggregating opinions of members who share information about investments and trading with other members of the platform. This aggregation of the traders' knowledge has led to the emergence of a new class of investors who trade not only based on their own knowledge about the market but also on that of the crowd's opinions. Thanks to the new startups, which can aggregate opinions and information shared by investors on social media using big data and machine learning, many more individual traders are privy to the crowd's opinions. This new trend is vividly illustrated by the GameStop trading mania discussed in Chapter 3. As we mentioned, innovative investment firms have started tracking Reddit and Twitter for the sentiment of individual traders.

Nowadays, experts and individual investors post regularly about their beliefs about stocks, either bullish or bearish, on various stock message

DOI: 10.1201/9781003196198-7

boards and social media outlets like Twitter. Many traders who follow these experts on social media then trade based on the tweets. For example, in August 2013, the famous investor Carl Icahn posted on Twitter to disclose his position in the tech giant, Apple, causing the market valuation of Apple to rise by $12.5 billion in a single trading day (Yahoo Finance, 2013). Besides experts, elected political leaders also tweet about the economy, which causes stock markets to react, providing advantages to traders who follow them. For example, when the President of the U.S. tweeted negatively about Amazon, its market capitalization was reduced by $6 billion within a few minutes (Abbruzzese, 2017). Well-established companies like Bloomberg and Thomson Reuters are now integrating data collected from Twitter with analytical tools to interpret them. For example, Bloomberg uses Twitter feeds to provide insights to their customers about the general sentiment of a stock (Garg et al., 2020). Elon Musk frequently tweeted about Bitcoin, causing the price of the cryptocurrency to change dramatically (Browne, 2021).

Another example of Fintech that is enabled by the aggregation of opinions is social trading and investment platforms. These platforms are the financial counterparts of social networks where people can create an online profile, see each other's profiles, and share information about investments and trading in financial instruments such as stocks and cryptocurrencies with other platform members. Members of a social trading platform range from novice to expert traders. Online platforms, like Sharewise and Estimize, specialize in aggregating the opinions and predictions of all its users to come up with a target price for a stock, thus enabling their users to gain market insights from public opinions.

Garg et al. (2020) analyzed how social trading platforms are upending financial markets. They find that the democratization of the availability of the crowd's opinions through these information technology tools has significantly reduced market efficiency because public information is over-weighted in individual investors' decisions. Correcting such inefficiency in the age of social media is a challenge for social media and financial market researchers and practitioners.

7.2 SOCIAL MEDIA TEXT, NETWORK, IMAGES, AND VIDEOS AS DATA

Another future trend is to dig deep into the various data formats contained in social media. Textual information, network graphs, images, and videos can all be considered as data in analytics. Recall that in Chapter 3, we discussed how to use unstructured biography texts to measure user similarity in the framework of Latent Dirichlet Allocation topic modeling (Lee et al., 2016). In Chapter 4, we described how to use movie keywords to construct thematic similarity between these two movies (Lee et al., 2018). Textual data becomes an essential part of social media analytics.

Network graphs can also be treated as data in social media analytics. Firms and organizations increasingly use real-time performance feedback in social media applications to evaluate employees. Petryk et al. (2019) visualized rating networks within organizations: Employees are nodes, and connections between nodes exist if an evaluation between the pair occurs. They find that different aspects of network embeddedness affect performance rating scores differently. In particular, a rater's positional embeddedness (measured by eigenvector centrality) is positively associated with the rating score he or she gives. A rater's structural embeddedness (measured by outdegree centrality) is negatively associated with the rating score he or she gives. Although network graphs are sometimes difficult to obtain, they are useful indicators for the underlying information transmission among individuals.

New image recognition methods and video analytics make images and videos easy to analyze. A few recent studies employ artificial intelligence (AI) and machine learning in image and video content analysis. For example, Duan et al. (2020) examine the impact of perceived trustworthiness on online crowdfunding success. They employ AI-based facial detection techniques to collect a sample of crowdfunding creators' front photos from Kickstarter and automatically construct a measure of facial trustworthiness. A challenge to use images and videos as data is that there is no standard measure for image and video content information.

7.3 REAL-TIME NATURE OF SOCIAL MEDIA IN OPERATIONS

Recent years have witnessed an unprecedented trend of using social media information to improve firms' operational decisions (Cui et al., 2018; Qiu et al., 2021). The real-time nature of social media is unique in the applications of operational problems (He et al., 2021; Rivera et al., 2021). To deliver real-time feedback to support employee development and rapid innovation, many companies replace traditional performance management with systems that enable frequent and continuous employee evaluation. These social media-like applications enable supervisors and employees to give, seek, and receive competency-based feedback using their computers, smartphones, or other devices. Rivera et al. (2021) find that relationship source (peer, subordinate, or supervisor) impacts real-time feedback: Feedback tends to be more important when it is from supervisors. In addition, positive real-time feedback has a stronger effect on future ratings than negative feedback. These findings highlight how companies could use information systems to create an innovative human resource operation that delivers flexibility and agility. Comparing the relative effectiveness between real-time feedback and traditional performance management is an ongoing challenge for researchers and practitioners.

7.4 PATH FORWARD

The digital technology underlying the concept of Industry 4.0 has automated traditional manufacturing and brought new disruptive changes to the economy (Wang et al., 2021). As a type of digital technology, social media analytics is becoming more and more integrated with other technologies, such as AI, IoT, Natural language processing (NLP), and 5G. Together with other digital technologies, social media analytics not only benefits existing businesses, but also accelerates the trend of creating new business models and brings disruptive changes. Benefiting from the increased connectivity and efficient operation capabilities brought by social media technology, individuals with limited resources can now participate in innovations and entrepreneurial activities through social media platforms and become social media celebrities.

As we have seen in earlier chapters, social media-based strategies are fundamentally different from traditional firm strategies without considering the impact of social media content. The explosive development of digital technologies is posing new challenges and opportunities to companies. For example, managing consumer emotion is crucial in customer service nowadays because consumer emotion is contagious on social media platforms. However, if service providers use too much emotion, it may backfire since it seems not sincere, just like that a false smile may seem manipulative (Wang et al., 2021).

In this book, we have seen the incredible power of social media when it is combined with the wisdom of crowds, platforms, machine learning, telecommunications, and Fintech. In the future, we expect that social media will be integrated with other emerging technologies more tightly and seamlessly.

The reason is that social media is more than a technology; it is a way of bringing people together. As Goethe said: "All theory, dear friend, is gray, but the golden tree of life springs ever green" (Ratcliffe, 2017). Social media is part of the golden tree of our daily lives.

REFERENCES

Abbruzzese, J. (2017). A single Donald Trump tweet just lost Amazon $6 billion in value. *Mashable*. Last accessed on February 17, 2021, https://mashable.com/2017/08/16/donald-trump-amazon-tweet/.

Browne, R. (2021). Bitcoin falls after Elon Musk tweets breakup meme. *CNBC news*. Last accessed on June 26, 2021, https://www.cnbc.com/2021/06/04/bitcoin-falls-after-elon-musk-tweets-breakup-meme.html.

Cui, R., Gallino, S., Moreno, A., & Zhang, D. J. (2018). The operational value of social media information. *Production and Operations Management*, *27*(10), 1749–1769.

Duan, Y., Hsieh, T. S., Wang, R. R., & Wang, Z. (2020). Entrepreneurs' facial trustworthiness, gender, and crowdfunding success. *Journal of Corporate Finance*, *64*, 101693.

Garg, A., Qiu, L., & Bandyopadhyay, S. (2020). Beauty contest and social value of Fintech: An economic analysis. Americas Conference on Information Systems 2020.

He, S., Qiu, L., & Cheng, X. (2021). Surge pricing and short-term wage elasticity of labor supply in real-time ridesharing markets. *MIS Quarterly,* Forthcoming.

Lee, G. M., Qiu, L., & Whinston, A. B. (2016). A friend like me: Modeling network formation in a location-based social network. *Journal of Management Information Systems*, *33*(4), 1008–1033.

Lee, S. Y., Qiu, L., & Whinston, A. (2018). Sentiment manipulation in online platforms: An analysis of movie tweets. *Production and Operations Management*, *27*(3), 393–416.

Petryk, M., Rivera, M., Bhattacharya, S., Qiu, L., & Kumar, S. (2019). How network embeddedness affects real-time performance feedback: An empirical investigation. Working paper, Available at SSRN: https://ssrn.com/abstract=3438032.

Qiu, L., Hong, Y. K., & Whinston, A. (2021). Special issue of production and operations management social technologies in operations. *Production and Operations Management*, *30*(4), 1190–1191.

Ratcliffe, S. (2017). *Oxford essential quotations*. Oxford University Press, Oxford, United Kingdom. Online version can be accessed at: https://www.oxfordreference.com/view/10.1093/acref/9780191843730.001.0001/q-oro-ed5-00004900.

Rivera, M., Qiu, L., Kumar, S., & Petrucci, T. (2021). Are traditional performance reviews outdated? An empirical analysis on continuous, real-time feedback in the workplace. *Information Systems Research*, *32*(2), 517–540.

Wang, X., Jiang, M., Han, W., & Qiu, L. (2021). Do emotions sell? Impact of emotional expressions on sales in space-sharing economy. *Production and Operations Management*, Forthcoming.

Yahoo Finance. (2013). Carl Icahn's multibillion-dollar tweet boosts Apple stock. Last accessed on February 17, 2021, https://finance.yahoo.com/blogs/the-exchange/carl-icahn-multibillion-dollar-tweet-boosts-apple-stock-205938760.html.